PRINCIPLES OF GEOLOGY

PLACE IN RETURN BOX to remove this checkout from your record.
TO AVOID FINES return on or before date due.

APR 2 8 1992

DATE DUE	DATE DUE	DATE DUE
DEC 1 5 2005		
12 0 6 6		

This is volume II of a facsimile of the first edition of Lyell's
Principles of Geology, to which we have added this page and a
new title page only. The frontispiece and the map facing page
304 are hand painted in the original and reproduced here in
black and white.

We thank the University of Chicago Library for permission to
reprint this facsimile from a copy of the book in the Department
of Special Collections at the Joseph Regenstein Library at the
University of Chicago.

The University of Chicago Press, Chicago 60637
The University of Chicago Press, Ltd., London
Originally published in 1832 by John Murray, London
University of Chicago Press edition 1991
Printed in the United States of America

99 98 97 96 95 94 93 92 91   6 5 4 3 2 1

Library of Congress Cataloging-in-Publication Data
(Revised for volume 2)

Lyell, Charles, Sir, 1797–1875.
  Principles of geology.
  Reprint. Originally published: London : J. Murray, 1830–1833.
  Includes bibliographical references and indexes.
  1. Geology. I. Title.
QE26.L956   1990   550   90-11008
ISBN 0-226-49796-8 (cloth)
ISBN 0-226-49797-6 (pbk.)

♾ The paper used in this publication meets the minimum
requirements of the American National Standard for Information
Sciences—Permanence of Paper for Printed Library Materials,
ANSI Z39.48-1984

PRINCIPLES

OF

GEOLOGY

First Edition

VOLUME II

CHARLES LYELL

The University of Chicago Press

Chicago and London

View of the Valle del Bove Etna.

1. Montagnola.   2. Torre del filosofo.   3. Highest Cone.   4. Lapra.   5. Finocchio.   6. Capra.   7. Cone of 1811.   8. Cone del astro.   9. Musarra.

PRINCIPLES

OF

GEOLOGY,

BEING

AN ATTEMPT TO EXPLAIN THE FORMER CHANGES OF THE EARTH'S SURFACE,

BY REFERENCE TO CAUSES NOW IN OPERATION.

BY

CHARLES LYELL, Esq., F.R.S.

FOR. SEC. TO THE GEOL. SOC., PROF. OF GEOL. TO KING'S COLL., LONDON.

" The inhabitants of the globe, like all the other parts of it, are subject to change. It is not only the individual that perishes, but whole species."
" A change in the animal kingdom seems to be part of the order of nature, and is visible in instances to which human power cannot have extended."
PLAYFAIR, *Illustrations of the Huttonian Theory*, § 413.

VOLUME THE SECOND.

LONDON:

JOHN MURRAY, ALBEMARLE-STREET.

MDCCCXXXII.

LONDON:
Printed by WILLIAM CLOWES,
Stamford Street.

TO

WILLIAM JOHN BRODERIP, Esq., B. A.

BARRISTER AT LAW,

F.R.S., F.L.S., ETC.,

VICE PRESIDENT OF THE GEOLOGICAL SOCIETY OF LONDON.

My Dear Friend,

In dedicating this volume to you, I am glad of an opportunity of acknowledging the kind interest which you have uniformly taken in the success of my labours, and the valuable assistance which you have afforded me in several departments of Natural History.

I am,

My Dear Friend,

Yours, very sincerely,

Charles Lyell.

London, December 8th, 1831.

PREFACE

THE author has found it impossible to compress into two volumes, according to his original plan, the wide range of subjects which must be discussed, in order fully to explain his views respecting the causes of geological phenomena. As it will, therefore, be necessary to extend the " Principles of Geology" to three volumes, he prefers the publication of the present part without delay, because it brings to a close one distinct branch of the inquiry, the study of which will be found absolutely essential to the understanding of the theories hereafter to be proposed. Considerable progress has already been made in the remainder of the work, which will shortly be laid before the public.

London, December 8th, 1831.

CONTENTS.

Vol. II.

CHAPTER IX.

CHAPTER X.

CHAPTER XI.

CONTENTS.

ERRATA.

Frontispiece, *for* Montagnola *read* Montagnuola.

Page 39, line 2 from the bottom, *for* excusively *read* exclusively.

131 — 14 — top — Hypnum — Sphagnum.

147 — 21 — top, *dele* of.

178 — 8 — top, *for* even *read* ever.

PRINCIPLES OF GEOLOGY.

CHAPTER I.

Changes of the Organic World now in progress—Division of the subject—Examination of the question, Whether Species have a real existence in Nature? —Importance of this question in Geology—Sketch of Lamarck's arguments in favour of the Transmutation of Species, and his conjectures respecting the Origin of existing Animals and Plants—His Theory of the transformation of the Orang Outang into the Human Species.

In our first volume we treated of the changes which have taken place in the inorganic world within the historical era, and we must next turn our attention to those now in progress in the animate creation. In examining this class of phenomena, we shall treat first of the vicissitudes to which *species* are subject, and afterwards consider the influence of the powers of vitality in modifying the surface of the earth and the material constituents of its crust.

The first of these divisions will lead us, among other topics, to inquire, first, whether species have a real and permanent existence in nature; or whether they are capable, as some naturalists pretend, of being indefinitely modified in the course of a long series of generations? Secondly, whether, if species have a real existence, the individuals composing them have been derived originally from many similar stocks, or each from one only, the descendants of which have spread themselves gradually from a particular point over the habitable lands and waters? Thirdly, how far the duration of each species of animal and plant is limited by its dependance on certain fluctuating and temporary conditions in the state of the animate and inanimate world? Fourthly, whether there be proofs of the successive extermination of species in the ordinary course of nature, and whether there be any reason for conjecturing that new animals and plants are created from time to time, to supply their place?

Before we can advance a step in our proposed inquiry, we must be able to define precisely the meaning which we attach to the term species. This is even more necessary in geology than in the ordinary studies of the naturalist ; for they who deny that such a thing as a species exists, concede nevertheless that a botanist or zoologist may reason as if the specific character were constant, because they confine their observations to a brief period of time. Just as the geographer, in constructing his maps from century to century, may proceed as if the apparent places of the fixed stars remained absolutely the same, and as if no alteration was brought about by the precession of the equinoxes, so it is said in the organic world, the stability of a species may be taken as absolute, if we do not extend our views beyond the narrow period of human history ; but let a sufficient number of centuries elapse, to allow of important revolutions in climate, physical geography, and other circumstances, and the characters, say they, of the descendants of common parents may deviate indefinitely from their original type.

Now, if these doctrines be tenable, we are at once presented with a principle of incessant change in the organic world, and no degree of dissimilarity in the plants and animals which may formerly have existed, and are found fossil, would entitle us to conclude that they may not have been the prototypes and progenitors of the species now living. Accordingly, M. Geoffroy St. Hilaire has declared his opinion, that there has been an uninterrupted succession in the animal kingdom effected by means of generation, from the earliest ages of the world up to the present day ; and that the ancient animals whose remains have been preserved in the strata, however different, may nevertheless have been the ancestors of those now in being. Although this notion is not generally received, we feel that we are not warranted in assuming the contrary, without fully explaining the data and reasoning by which we conceive it may be refuted.

We shall begin by stating as concisely as possible all the facts and ingenious arguments by which the theory has been supported, and for this purpose we cannot do better than offer the reader a rapid sketch of Lamarck's statement of the proofs which he regards as confirmatory of the doctrine, and which he has derived partly from the works of his predecessors, and in part from original investigations.

We shall consider his proofs and inferences in the order in which they appear to have influenced his mind, and point out some of the results to which he was led while boldly following out his principles to their legitimate consequences.

The name of species, observes Lamarck, has been usually applied to ' every collection of similar individuals, produced by other individuals like themselves *.' This definition, he admits, is correct, because every living individual bears a very close resemblance to those from which it springs. But this is not all which is usually implied by the term species, for the majority of naturalists agree with Linnæus in supposing that all the individuals propagated from one stock have certain distinguishing characters in common which will never vary, and which have remained the same since the creation of each species.

In order to shake this opinion, Lamarck enters upon the following line of argument. The more we advance in the knowledge of the different organized bodies which cover the surface of the globe, the more our embarrassment increases, to determine what ought to be regarded as a species, and still more how to limit and distinguish genera. In proportion as our collections are enriched, we see almost every void filled up, and all our lines of separation effaced ; we are reduced to arbitrary determinations, and are sometimes fain to seize upon

* Phil. Zool. tom. i. p.54.

the slight differences of mere varieties, in order to form characters for what we choose to call a species, and sometimes we are induced to pronounce individuals but slightly differing, and which others regard as true species, to be varieties.

The greater the abundance of natural objects assembled together, the more do we discover proofs that everything passes by insensible shades into something else; that even the more remarkable differences are evanescent, and that nature has, for the most part, left us nothing at our disposal for establishing distinctions, save trifling and, in some respects, puerile particularities.

We find that many genera amongst animals and plants are of such an extent, in consequence of the number of species referred to them, that the study and determination of these last has become almost impracticable. When the species are arranged in a series, and placed near to each other, with due regard to their natural affinities, they each differ in so minute a degree from those next adjoining, that they almost melt into each other, and are in a manner confounded together. If we see isolated species, we may presume the absence of some more closely connected, and which have not yet been discovered. Already are there genera, and even entire orders,—nay, whole classes, which present an approximation to the state of things here indicated.

If, when species have been thus placed in a regular series, we select one, and then, making a leap over several intermediate ones, we take a second, at some distance from the first, these two will, on comparison, be seen to be very dissimilar; and it is in this manner that every naturalist begins to study the objects which are at his own door. He then finds it an easy task to establish generic and specific distinctions; and it is only when his experience is enlarged, and when he has made himself master of the intermediate links, that his difficulties and ambiguities begin. But while we are thus compelled to resort to trifling and minute characters in our attempt to separate species, we find a striking disparity between individuals which we know to have descended from a common stock, and these newly-acquired peculiarities are regularly transmitted from one generation to another, constituting what are called *races.*

From a great number of facts, continues the author, we learn that, in proportion as the individuals of one of our species change their situation, climate, and manner of living, they change also, by little and little, the consistence and proportions of their parts, their form, their faculties, and even their organization, in such a manner, that everything in them comes at last to participate in the mutations to which they have been exposed. Even in the same climate a great difference of situation and exposure causes individuals to vary ; but if these individuals continue to live and to be reproduced under the same difference of circumstances, distinctions are brought about in them which become in some degree essential to their existence. In a word, at the end of many successive generations, these individuals, which originally belonged to another species, are transformed into a new and distinct species *.

Thus, for example, if the seeds of a grass, or any other plant which grows naturally in a moist meadow, be accidentally transported, first to the slope of some neighbouring hill, where the soil, although at a greater elevation, is damp enough to allow the plant to live ; and if, after having lived there, and having been several times regenerated, it reaches by degrees the drier and almost arid soil of a mountain declivity, it will then, if it succeeds in growing and perpetuates itself for a series of generations, be so changed that botanists who meet with it will regard it as a particular species†. The unfavourable climate in this case,. deficiency of nourishment, exposure to the winds, and other causes, give rise to a stunted and dwarfish race, with some organs more developed than others, and having proportions often quite peculiar.

* Phil. Zool. tom. i. p. 62.   † Ibid.

What nature brings about in a great lapse of time we occasion suddenly by changing the circumstances in which a species has been accustomed to live. All are aware that vegetables taken from their birth-place and cultivated in gardens, undergo changes which render them no longer recognizable as the same plants. Many which were naturally hairy become smooth or nearly so; a great number of such as were creepers and trailed along the ground, rear their stalks and grow erect. Others lose their thorns or asperities; others again, from the ligneous state which their stem possessed in hot climates, where they were indigenous, pass to the herbaceous, and, among them, some which were perennials become mere annuals. So well do botanists know the effects of such changes of circumstances, that they are averse to describe species from garden specimens, unless they are sure that they have been cultivated for a very short period.

‘ Is not the cultivated wheat,’ (*Triticum sativum*) asks Lamarck, ‘ a vegetable brought by man into the state in which ‘ we now see it ? Let any one tell me in what country a similar ‘ plant grows wild, unless where it has escaped from cultivated ‘ fields ? Where do we find in nature our cabbages, lettuces, ‘ and other culinary vegetables, in the state in which they ‘ appear in our gardens ? Is it not the same in regard to ‘ a great quantity of animals which domesticity has changed ‘ or considerably modified * ?’ Our domestic fowls and pigeons are unlike any wild birds. Our domestic ducks and geese have lost the faculty of raising themselves into the higher regions of the air, and crossing extensive countries in their flight, like the wild ducks and wild geese from which they were originally derived. A bird which we breed in a cage cannot, when restored to liberty, fly like others of the same species which have been always free. This small alteration of circumstances, however, has only diminished the power of flight, without modifying the form of any part of the wings. But when individuals of the same race are retained in captivity

* Phil. Zool. tom. i. p. 227.

during a considerable length of time, the form even of their parts is gradually made to differ, especially if climate, nourishment, and other circumstances, be also altered.

The numerous races of dogs which we have produced by domesticity are nowhere to be found in a wild state. In nature we should seek in vain for mastiffs, harriers, spaniels, greyhounds, and other races, between which the differences are sometimes so great, that they would be readily admitted as specific between wild animals ; ' yet all these have sprung ' originally from a single race, at first approaching very near ' to a wolf, if, indeed, the wolf be not the true type which at ' some period or other was domesticated by man.'

Although important changes in the nature of the places which they inhabit modify the organization of animals as well as vegetables, yet the former, says Lamarck, require more time to complete a considerable degree of transmutation, and, consequently, we are less sensible of such occurrences. Next to a diversity of the medium in which animals or plants may live, the circumstances which have most influence in modifying their organs are differences in exposure, climate, the nature of the soil, and other local particulars. These *circumstances* are as varied as are the characters of species, and, like them, pass by insensible shades into each other, there being every intermediate gradation between the opposite extremes. But each locality remains for a very long time the same, and is altered so slowly that we can only become conscious of the reality of the change, by consulting geological monuments, by which we learn that the order of things which now reigns in each place has not always prevailed, and by inference anticipate that it will not always continue the same *.

Every considerable alteration in the local circumstances in which each race of animals exists, causes a change in their wants, and these new wants excite them to new actions and habits. These actions require the more frequent employment of some parts before but slightly exercised, and then greater develop-

* Phil. Zool. tom. i. p. 232.

ment follows as a consequence of their more frequent use. Other organs no longer in use are impoverished and diminished in size, nay, are sometimes entirely annihilated, while in their place new parts are insensibly produced for the discharge of new functions*.

We must here interrupt the author's argument, by observing that no positive fact is cited to exemplify the substitution of some *entirely new* sense, faculty, or organ, in the room of some other suppressed as useless. All the instances adduced go only to prove that the dimensions and strength of members and the perfection of certain attributes may, in a long succession of generations, be lessened and enfeebled by disuse; or, on the contrary, be matured and augmented by active exertion, just as we know that the power of scent is feeble in the greyhound, while its swiftness of pace and its acuteness of sight are remarkable—that the harrier and stag-hound, on the contrary, are comparatively slow in their movements, but excel in the sense of smelling.

We point out to the reader this important chasm in the chain of the evidence, because he might otherwise imagine that we had merely omitted the illustrations for the sake of brevity, but the plain truth is, that there were no examples to be found ; and when Lamarck talks ' of the efforts of internal sentiment,' ' the influence of subtle fluids,' and the ' acts of organization,' as causes whereby animals and plants may acquire *new organs*, he gives us names for things, and with a disregard to the strict rules of induction, resorts to fictions, as ideal as the ' plastic virtue,' and other phantoms of the middle ages.

It is evident, that if some well authenticated facts could have been adduced to establish one complete step in the process of transformation, such as the appearance, in individuals descending from a common stock, of a sense or organ entirely new, and a complete disappearance of some other enjoyed by their progenitors, that time alone might then be supposed

* Phil. Zool. tom. i. p. 234.

sufficient to bring about any amount of metamorphosis. The gratuitous assumption, therefore, of a point so vital to the theory of transmutation, was unpardonable on the part of its advocate.

But to proceed with the system; it being assumed as an undoubted fact, that a change of external circumstances may cause one organ to become entirely obsolete, and a new one to be developed such as never before belonged to the species, the following proposition is announced, which, however staggering and absurd it may seem, is logically deduced from the assumed premises. " It is not the organs, or, in other words, the nature and form of the parts of the body of an animal which have given rise to its habits, and its particular faculties, but on the contrary, its habits, its manner of living, and those of its progenitors have in the course of time determined the form of its body, the number and condition of its organs, in short, the faculties which it enjoys. Thus otters, beavers, water-fowl, turtles, and frogs, were not made web-footed in order that they might swim ; but their wants having attracted them to the water in search of prey, they stretched out the toes of their feet to strike the water and move rapidly along its surface. By the repeated stretching of their toes, the skin which united them at the base acquired a habit of extension, until in the course of time the broad membranes which now connect their extremities were formed.

In like manner the antelope and the gazelle were not endowed with light agile forms, in order that they might escape by flight from carnivorous animals ; but having been exposed to the danger of being devoured by lions, tigers, and other beasts of prey, they were compelled to exert themselves in running with great celerity, a habit which, in the course of many generations, gave rise to the peculiar slenderness of their legs, and the agility and elegance of their forms.

The cameleopard was not gifted with a long flexible neck because it was destined to live in the interior of Africa, where the soil was arid and devoid of herbage, but being reduced by the nature of that country to support itself on the foliage of lofty trees, it contracted a habit of stretching itself up to reach the high boughs, until its fore-legs became longer than the hinder, and its neck so elongated, that it could raise its head to the height of twenty feet above the ground."

Another line of argument is then entered upon, in farther corroboration of the instability of species. In order it is said that individuals should perpetuate themselves unaltered by generation, those belonging to one species ought never to ally themselves to those of another : but such sexual unions do take place, both among plants and animals; and although the offspring of such irregular connexions are usually sterile, yet such is not always the case. Hybrids have sometimes proved prolific where the disparity between the species was not too great; and by this means alone, says Lamarck, varieties may gradually be created by near alliances, which would become races, and in the course of time would constitute what we term species*.

But if the soundness of all these arguments and inferences be admitted, we are next to inquire, what were the original types of form, organization, and instinct, from which the diversities of character, as now exhibited by animals and plants, have been derived? We know that individuals which are mere varieties of the same species, would, if their pedigree could be traced back far enough, terminate in a single stock ; so according to the train of reasoning before described, the species of a genus, and even the genera of a great family, must have had a common point of departure. What then was the single stem from which so many varieties of form have rami-fied? Were there many of these, or are we to refer the origin of the whole animate creation, as the Egyptian priests did that of the universe, to a single egg?

In the absence of any positive data for framing a theory on so obscure a subject, the following considerations were deemed of importance to guide conjecture.

* Phil. Zool p. 64.

In the first place, if we examine the whole series of known animals, from one extremity to the other, when they are arranged in the order of their natural relations, we find that we may pass progressively, or at least with very few interruptions, from beings of more simple to those of a more compound structure ; and in proportion as the complexity of their organization increases, the number and dignity of their faculties increase also. Among plants a similar approximation to a graduated scale of being is apparent. Secondly, it appears from geological observations, that plants and animals of more simple organization existed on the globe before the appearance of those of more compound structure, and the latter were successively formed at later periods : each new race being more fully developed than the most perfect of the preceding era.

Of the truth of the last-mentioned geological theory, Lamarck seems to have been fully persuaded ; and he also shews that he was deeply impressed with a belief prevalent amongst the older naturalists, that the primeval ocean invested the whole planet long after it became the habitation of living beings, and thus he was inclined to assert the priority of the types of marine animals to those of the terrestrial, and to fancy, for example, that the testacea of the ocean existed first, until some of them, by gradual evolution, were *improved* into those inhabiting the land.

These speculative views had already been, in a great degree, anticipated by Delamétherie in his Teliamed, and by several modern writers, so that the tables were completely turned on the philosophers of antiquity, with whom it was a received maxim, that created things were always most perfect when they came first from the hands of their Maker, and that there was a tendency to progressive deterioration in sublunary things when left to themselves—

———————— omnia fatis
In pejus ruere, ac retrò sublapsa referri.

So deeply was the faith of the ancient schools of philosophy imbued with this doctrine, that to check this universal proneness to degeneracy, nothing less than the re-intervention of the Deity was thought adequate; and it was held, that thereby the order, excellence, and pristine energy of the moral and physical world had been repeatedly restored.

But when the possibility of the indefinite modification of individuals descending from common parents was once assumed, as also the geological generalization respecting the progressive development of organic life, it was natural that the ancient dogma should be rejected, or rather reversed; and that the most simple and imperfect forms and faculties should be conceived to have been the originals whence all others were developed. Accordingly, in conformity to these views, inert matter was supposed to have been first endowed with life; until in the course of ages, sensation was superadded to mere vitality : sight, hearing, and the other senses, were afterwards acquired ; and then instinct and the mental faculties; until, finally, by virtue of the tendency of things to *progressive improvement*, the irrational was developed into the rational.

The reader, however, will immediately perceive, that if all the higher orders of plants and animals were thus supposed to be comparatively modern, and to have been derived in a long series of generations from those of more simple conformation, some further hypothesis became indispensable, in order to explain why, after an indefinite lapse of ages, there were still so many beings of the simplest structure. Why have the majority of existing creatures remained stationary throughout this long succession of epochs, while others have made such prodigious advances? Why are there still such multitudes of infusoria and polypes, or of confervæ and other cryptogamic plants ? Why, moreover, has the process of development acted with such unequal and irregular force on those classes of beings which have been greatly perfected, so that there are wide chasms in the series; gaps so enormous, that Lamarck fairly admits we can never expect to fill them up by future discoveries ?

The following hypothesis was provided to meet these objections. Nature, we are told, is not an intelligence, nor the Deity, but a delegated power—a mere instrument—a piece of mechanism acting by necessity—an order of things constituted by the Supreme Being, and subject to laws which are the expressions of his will. This nature is *obliged* to proceed gradually in all her operations; she cannot produce animals and plants of all classes at once, but must always begin by the formation of the most simple kinds; and out of them elaborate the more compound, adding to them successively, different systems of organs, and multiplying more and more their number and energy.

This Nature is daily engaged in the formation of the elementary rudiments of animal and vegetable existence, which correspond to what the ancients termed *spontaneous generations*. She is always beginning anew, day by day, the work of creation, by forming monads, or ' rough draughts' (ébauches), which are the only living things she ever gives birth to *directly*.

There are distinct primary rudiments of plants and animals, and *probably* of each of the great divisions of the animal and vegetable kingdoms *. These are gradually developed into the higher and more perfect classes by the slow, but unceasing agency of two influential principles : first, *the tendency to progressive advancement* in organization, accompanied by greater dignity in instinct, intelligence, &c.; secondly, *the force of external circumstances*, or of variations in the physical condition of the earth, or the mutual relations of plants and animals. For as species spread themselves gradually over the globe, they are exposed from time to time to variations in climate, and to changes in the quantity and quality of their food ; they meet with new plants and animals which assist or retard their development, by supplying them with nutriment, or destroying their foes. The nature also of each locality is in itself fluctuating, so that even if the relation of other animals and plants were invariable, the habits and organization

* Animaux sans Vert., tom. i. p. 56, Introduction.

of species would be modified by the influence of local revolutions.

Now, if the first of these principles, *the tendency to progressive development,* were left to exert itself with perfect freedom, it would give rise, says Lamarck, in the course of ages, to a graduated scale of being, where the most insensible transition might be traced from the simplest to the most compound structure, from the humblest to the most exalted degree of intelligence. But in consequence of the perpetual interference of the *external causes* before mentioned, this regular order is greatly interfered with, and an approximation only to such a state of things is exhibited by the animate creation, the progress of some races being retarded by unfavourable, and that of others accelerated by favourable, combinations of circumstances. Hence, all kinds of anomalies interrupt the continuity of the plan, and chasms, into which whole genera or families might be inserted, are seen to separate the nearest existing portions of the series.

Such is the machinery of the Lamarckian system ; but our readers will hardly, perhaps, be able to form a perfect conception of so complicated a piece of mechanism, unless we exhibit it in motion, and shew in what manner it can work out, under the author's guidance, all the extraordinary effects which we behold in the present state of the animate creation. We have only space for exhibiting a small part of the entire process by which a complete metamorphosis is achieved, and shall, therefore, omit the mode whereby, after a countless succession of generations, a small gelatinous body is transformed into an oak or an ape. We pass on at once to the last grand step in the progressive scheme, whereby the orang-outang, having been already evolved out of a monad, is made slowly to attain the attributes and dignity of man.

One of the races of quadrumanous animals which had reached the highest state of perfection, lost, by constraint of circumstances, (concerning the exact nature of which tradition is unfortunately silent,) the habit of climbing trees, and of hanging on by grasping the boughs with their feet as with hands. The individuals of this race being obliged for a long series of generations to use their feet exclusively for walking, and ceasing to employ their hands as feet, were transformed into bimanous animals, and what before were thumbs became mere toes, no separation being required when their feet were used solely for walking. Having acquired a habit of holding themselves upright, their legs and feet assumed insensibly a conformation fitted to support them in an erect attitude, till at last these animals could no longer go on all fours without much inconvenience.

The Angola orang, *Simia troglodytes*, Linn., is the most perfect of animals, much more so than the Indian orang, *Simia Satyrus*, which has been called the orang-outang, although *both* are *very inferior* to man in corporeal powers and intelligence. These animals frequently hold themselves upright, but their organization has *not yet* been sufficiently modified to sustain them habitually in this attitude, so that the standing posture is very uneasy to them. When the Indian orang is compelled to take flight from pressing danger, he immediately falls down upon all fours, shewing clearly that this was the original position of the animal. Even in man, whose organization, in the course of a long series of generations, has advanced so much farther, the upright posture is fatiguing and can only be supported for a limited time, and by aid of the contraction of many muscles. If the vertebral column formed the axis of the human body, and supported the head and all the other parts in equilibrium, then might the upright position be a state of repose ; but as the human head does not articulate in the centre of gravity; as the chest, belly, and other parts, press almost entirely forward with their whole weight, and as the vertebral column reposes upon an oblique base, a watchful activity is required to prevent the body from falling. Children which have large heads and prominent bellies can hardly walk at the end even of two years, and their frequent tumbles indicate the natural tendency in man to resume the quadrupedal state.

Now, when so much progress had been made by the quadrumanous animals before mentioned, that they could hold themselves habitually in an erect attitude, and were accustomed to a wide range of vision, and ceased to use their jaws for fighting, and tearing, or for clipping herbs for food, their snout became gradually shorter, their incisor teeth became vertical, and the facial angle grew more open.

Among other ideas which the natural *tendency to perfection* engendered, the desire of ruling suggested itself, and this race succeeded at length in getting the better of the other animals, and made themselves masters of all those spots on the surface of the globe which best suited them. They drove out the animals which approached nearest to them in organization and intelligence, and which were in a condition to dispute with them the good things of this world, forcing them to take refuge in deserts, woods and wildernesses, where their multiplication was checked, and the progressive development of their faculties retarded, while in the mean time the dominant race spread itself in every direction, and lived in large companies where new wants were successively created, exciting them to industry, and gradually perfecting their means and faculties.

In the supremacy and increased intelligence acquired by the ruling race, we see an illustration of the natural tendency of the organic world to grow more perfect, and in their influence in repressing the advance of others, an example of one of those disturbing causes before enumerated, that *force of external circumstances*, which causes such wide chasms in the regular series of animated beings.

When the individuals of the dominant race became very numerous, their ideas greatly increased in number, and they felt the necessity of communicating them to each other, and of augmenting and varying the signs proper for the communication of ideas. Meanwhile the inferior quadrumanous animals, although most of them were gregarious, acquired no new ideas, being persecuted and restless in the deserts, and obliged to fly and conceal themselves, so that they conceived no new wants. Such ideas as they already had remained unaltered, and they could dispense with the communication of the greater part of these. To make themselves, therefore, understood by their fellows, required merely a few movements of the body or limbs—whistling, and the uttering of certain cries varied by the inflexions of the voice.

On the contrary, the individuals of the ascendant race, animated with a desire of interchanging their ideas, which became more and more numerous, were prompted to multiply the means of communication, and were no longer satisfied with mere pantomimic signs, nor even with all the possible inflexions of the voice, but made continual efforts to acquire the power of uttering articulate sounds, employing a few at first, but afterwards varying and perfecting them according to the increase of their wants. The habitual exercise of their throat, tongue and lips, insensibly modified the conformation of these organs, until they became fitted for the faculty of speech *.

In effecting this mighty change, " the exigencies of the individuals were the sole agents, they gave rise to efforts, and the organs proper for articulating sounds were developed by their habitual employment." Hence, in this peculiar race, the origin of the admirable faculty of speech ; hence also the diversity of languages, since the distance of places where the individuals composing the race established themselves, soon favoured the corruption of conventional signs †.

* Lamarck's Phil. Zool., tom. i. p. 356.   † Ibid. p. 357.

CHAPTER II

Recapitulation of the arguments in favour of the theory of transmutation of species —Their insufficiency—The difficulty of discriminating species mainly attributable to a defective knowledge of their history—Some mere varieties possibly more distinct than certain individuals of distinct species—Variability in a species consistent with a belief that the limits of deviation are fixed—No facts of transmutation authenticated—Varieties of the Dog—The Dog and Wolf distinct species—Mummies of various animals from Egypt identical in character with living individuals—Seeds and plants from the Egyptian tombs—Modifications produced in plants by agriculture and gardening.

THE theory of the transmutation of species, considered in the last chapter, has met with some degree of favour from many naturalists, from their desire to dispense, as far as possible, with the repeated intervention of a First Cause, as often as geological monuments attest the successive appearance of new races of animals and plants, and the extinction of those pre-existing. But, independently of a predisposition to account, if possible, for a series of changes in the organic world, by the regular action of secondary causes, we have seen that many perplexing difficulties present themselves to one who attempts to establish the nature and the reality of the specific character. And if once there appears ground of reasonable doubt, in regard to the constancy of species, the amount of transformation which they are capable of undergoing, may seem to resolve itself into a mere question of the quantity of time assigned to the past duration of animate existence.

Before we enter upon our reasons for rejecting Lamarck's hypothesis, we shall recapitulate, in a few words, the phenomena, and the whole train of thought, by which we conceive it to have been suggested, and which have gained for this and analogous theories, both in ancient and modern times, a considerable number of votaries.

In the first place, the various groups into which plants and animals may be thrown, seem almost invariably, to a beginner, to be so natural, that he is usually convinced at first, as was Linnæus to the last, " that genera are as much founded in nature as the species which compose them *." When, by examining the numerous intermediate gradations, the student finds all lines of demarcation to be in most instances obliterated, even where they at first appeared most distinct, he grows more and more sceptical as to the real existence of genera, and finally regards them as mere arbitrary and artificial signs, invented like those which serve to distinguish the heavenly constellations for the convenience of classification, and having as little pretensions to reality.

Doubts are then engendered in his mind as to whether species may not also be equally unreal. The student is probably first struck with the phenomenon, that some individuals are made to deviate widely from the ordinary type by the force of peculiar circumstances, and with the still more extraordinary fact, that the newly-acquired peculiarities are faithfully transmitted to the offspring. How far, he asks, may such variations extend in the course of indefinite periods of time, and during great vicissitudes in the physical condition of the globe? His growing incertitude is at first checked by the reflection, that nature has forbidden the intermixture of the descendants of distinct original stocks, or has, at least, entailed sterility on their offspring, thereby preventing their being confounded together, and pointing out that a multitude of distinct types must have been created in the beginning, and must have remained pure and uncorrupted to this day.

Relying on this general law, he endeavours to solve each difcult problem by direct experiment, until he is again astounded by the phenomenon of a prolific hybrid, and still more by an example of a hybrid perpetuating itself throughout several generations in the vegetable world. He then feels himself reduced to the dilemma of choosing between two alternatives, either to reject the test, or to declare that the two species, from the union of

* Sir J. Smith's Introduction to Botany.

which the fruitful progeny has sprung, were mere varieties. If he prefer the latter, he is compelled to question the reality of the distinctness of all other supposed species which differ no more than the parents of such prolific hybrids ; for although he may not be enabled immediately to procure, in all such instances, a fruitful offspring, yet experiments show, that after repeated failures the union of two recognized species may at last, under very favourable circumstances, give birth to a fertile progeny.   Such circumstances, therefore, the naturalist may conceive to have occurred again and again, in the course of a great lapse of ages.

His first opinions are now fairly unsettled, and every stay at which he has caught has given way one after another ; he is in danger of falling into any new and visionary doctrine which may be presented to him ; for he now regards every part of the animate creation as void of stability, and in a state of continual flux.   In this mood he encounters the Geologist, who relates to him how there have been endless vicissitudes in the shape and structure of organic beings in former ages—how the approach to the present system of things has been gradual— that there has been a progressive development of organization subservient to the purposes of life, from the most simple to the most complex state—that the appearance of man is the last phenomenon in a long succession of events—and, finally, that a series of physical revolutions can be traced in the inorganic world, coeval and coextensive with those of organic nature.

These views seem immediately to confirm all his precon-ceived doubts as to the stability of the specific character, and he thinks he can discern an inseparable connexion between a series of changes in the inanimate world, and the capability of species to be indefinitely modified by the influence of external circum-stances.   Henceforth his speculations know no definite bounds ; he gives the rein to conjecture, and fancies that the outward form, internal structure, instinctive faculties, nay, that reason itself, may have been gradually developed from some of the simplest states of existence,—that all animals, that man himself, and the irrational beings, may have had one common origin; that all may be parts of one continuous and progressive scheme of development from the most imperfect to the more complex; in fine, he renounces his belief in the high genealogy of his species, and looks forward, as if in compensation, to the future perfectibility of man in his physical, intellectual, and moral attributes.

Let us now proceed to consider what is defective in evidence, and what fallacious in reasoning, in the grounds of these strange conclusions. Blumenbach judiciously observes, " that no general rule can be laid down for determining the distinctness of species, as there is no particular class of characters which can serve as a criterion. In each case we must be guided by *analogy* and *probability*." The multitude, in fact, and complexity of the proofs to be weighed, is so great, that we can only hope to obtain presumptive evidence, and we must, therefore, be the more careful to derive our general views as much as possible from those observations where the chances of deception are least. We must be on our guard not to tread in the footsteps of the naturalists of the middle ages, who believed the doctrine of spontaneous generation to be applicable to all those parts of the animal and vegetable kingdoms which they least understood, in direct contradiction to the analogy of all the parts best known to them; and who, when at length they found that insects and cryptogamous plants were also propagated from eggs and seeds, still persisted in retaining their old prejudices respecting the infusory animalcules and other minute beings, the generation of which had not then been demonstrated by the microscope to be governed by the same laws.

Lamarck has indeed attempted to raise an argument in favour of his system, out of the very confusion which has arisen in the study of some orders of animals and plants, in consequence of the slight shades of difference which separate the new species discovered within the last half century. That the embarrassment of those who attempt to classify and distinguish the new acquisitions poured in such multitudes into our museums, should increase with the augmentation of their number is quite natural; for to obviate this it is not enough that our powers of discrimination should keep pace with the increase of the objects, but we ought to possess greater opportunities of studying each animal and plant in all stages of its growth, and to know profoundly their history, their habits and physiological characters, throughout several generations. For, in proportion as the series of known animals grows more complete, none can doubt that there is a nearer approximation to a graduated scale of being; and thus the most closely allied species will possess a greater number of characters in common.

But, in point of fact, our new acquisitions consist, more and more as we advance, of specimens brought from foreign and often very distant and barbarous countries. A large proportion have never even been seen alive by scientific inquirers. Instead of having specimens of the young, the adult, and the aged individuals of each sex, and possessing means of investigating the anatomical structure, the peculiar habits and instincts of each, what is usually the state of our information? A single specimen, perhaps, of a dried plant, or a stuffed bird or quadruped; a shell without the soft parts of the animal; an insect in one stage of its numerous transformations; these are the scanty and imperfect data, which the naturalist possesses. Such information may enable us to separate species which stand at a considerable distance from each other; but we have no right to expect anything but difficulty and ambiguity, if we attempt, from such imperfect opportunities, to obtain distinctive marks for defining the characters of species, which are closely related.

If Lamarck could introduce so much certainty and precision into the classification of several thousand species of recent and fossil shells, notwithstanding the extreme remoteness of the organization of these animals from the type of those vertebrated species which are best known, and in the absence of so many of the living inhabitants of shells, we are led to form an exalted conception of the degree of exactness to which specific distinctions are capable of being carried, rather than to call in question their reality.

When our data are so defective, the most acute naturalist must expect to be sometimes at fault, and, like the novice, to overlook essential points of difference, passing unconsciously from one species to another, until, like one who is borne along in a current, he is astonished, on looking back, at observing that he has reached a point so remote from that whence he set out.

It is by no means improbable that when the series of species of certain genera is very full, they may be found to differ less widely from each other, than do the mere varieties or races of certain species. If such a fact could be established, it would by no means overthrow our confidence in the reality of species, although it would certainly diminish the chance of our obtaining certainty in our results.

It is almost necessary, indeed, to suppose, that varieties will differ in some cases, more decidedly than some species, if we admit that there is a graduated scale of being, and assume that the following laws prevail in the economy of the animate creation:—first, that the organization of individuals is capable of being modified to a limited extent by the force of external causes; secondly, that these modifications are, to a certain extent, transmissible to their offspring; thirdly, that there are fixed limits beyond which the descendants from common parents can never deviate from a certain type; fourthly, that each species springs from one original stock, and can never be permanently confounded, by intermixing with the progeny of any other stock; fifthly, that each species shall endure for a considerable period of time. Now if we assume, for the present, these rules hypothetically, let us see what consequences may naturally be expected to result.

We must suppose, that when the Author of Nature creates an animal or plant, all the possible circumstances in which its descendants are destined to live are foreseen, and that an organization is conferred upon it which will enable the species to perpetuate itself and survive under all the varying circumstances to which it must be inevitably exposed. Now the range of variation of circumstances will differ essentially in almost every case. Let us take for example any one of the most influential conditions of existence, such as temperature. In some extensive districts near the equator, the thermometer might never vary throughout several thousand centuries for more than 20° Fahrenheit; so that if a plant or animal be provided with an organization fitting it to endure such a range, it may continue on the globe for that immense period, although every individual might be liable at once to be cut off by the least possible excess of heat or cold beyond the determinate quantity. But if a species be placed in one of the temperate zones, and have a constitution conferred on it capable of supporting a similar range of temperature only, it will inevitably perish before a single year has passed away.

The same remark might be applied to any other condition, as food for example; it may be foreseen that the supply will be regular throughout indefinite periods in one part of the world, and in another very precarious and fluctuating both in kind and quantity. Different qualifications may be required for enabling species to live for a considerable time under circumstances so changeable. If, then, temperature and food be among those external causes, which according to certain laws of animal and vegetable physiology modify the organization, form, or faculties of individuals, we instantly perceive that the degrees of variability from a common standard must differ widely in the two cases above supposed, since there is a necessity of accommodating a species in one case to a much greater latitude of circumstances than in the other.

If it be a law, for instance, that scanty sustenance should check those individuals in their growth which are enabled to accommodate themselves to privations of this kind, and that a parent prevented in this manner from attaining the size proper to its species should produce a dwarfish offspring, a stunted race will arise, as is remarkably exemplified in some varieties of the horse and dog. The difference of stature in some races of dogs in comparison to others, is as one to five in linear dimensions, making a difference of a hundred-fold in volume*. Now there is good reason to believe that species in general are by no means susceptible of existing under a diversity of circumstances, which may give rise to such a disparity in size, and consequently, there will be a multitude of distinct species, of which no two adult individuals can ever depart so widely from a certain standard of dimensions as the mere varieties of certain other species,—the dog for instance. Now we have only to suppose that what is true of size, may also hold in regard to colour and many other attributes, and it will at once follow that the degree of possible discordance between varieties of the same species, may in certain cases exceed the utmost disparity which can even arise between two individuals of many distinct species.

The same remarks may hold true in regard to instincts; for if it be foreseen that one species will have to encounter a great variety of foes, it may be necessary to arm it with great cunning and circumspection, or with courage or other qualities capable of developing themselves on certain occasions; such for example as those migratory instincts which are so remarkably exhibited at particular periods, after they have remained dormant for many generations. The history and habits of one variety of such a species, may often differ more considerably from some other than those of many distinct species which have no such latitude of accommodation to circumstances.

Lamarck has somewhat misstated the idea commonly entertained of a species, for it is not true that naturalists in general assume that the organization of an animal or plant remains absolutely constant, and that it can never vary in any of its parts. All must be aware that circumstances influence the habits, and that the habits may alter the state of the parts and

* Cuvier, Disc. Prelim., p. 128, sixth edition.

organs *. But the difference of opinion relates to the extent
to which these modifications of the habits and organs of a
particular species may be carried.

Now let us first inquire what positive facts can be adduced
in the history of known species, to establish a great and permanent amount of change in the form, structure, or instinct
of individuals descending from some common stock. The
best authenticated examples of the extent to which species can
be made to vary, may be looked for in the history of domesticated animals and cultivated plants. It usually happens that
those species, both of the animal and vegetable kingdom, which
have the greatest pliability of organization, those which are
most capable of accommodating themselves to a great variety
of new circumstances, are most serviceable to man. These
only can be carried by him into different climates, and can have
their properties or instincts variously diversified by differences
of nourishment and habits. If the resources of a species be so
limited, and its habits and faculties be of such a confined and
local character, that it can only flourish in a few particular
spots, it can rarely be of great utility.

We may consider, therefore, that in perfecting the arts of
domesticating animals and cultivating plants, mankind have first
selected those species which have the most flexible frames and
constitutions, and have then been engaged for ages in conducting a series of experiments, with much patience and at great cost,
to ascertain what may be the greatest possible deviation from a
common type which can be elicited in these extreme cases.

The modifications produced in the different races of dogs,
exhibit the influence of man in the most striking point of view.
These animals have been transported into every climate, and
placed in every variety of circumstances; they have been made,
as a modern naturalist observes, the servant, the companion,
the guardian, and the intimate friend of man, and the power of
a superior genius has had a wonderful influence, not only on

* Phil. Zool., tom. i. p. 266.

their forms, but on their manners and intelligence *. Different races have undergone remarkable changes in the quantity and colour of their clothing: the dogs of Guinea are almost naked, while those of the Arctic circle are covered with a warm coat both of hair and wool, which enables them to bear the most intense cold without inconvenience. There are differences also of another kind no less remarkable, as in size, the length of their muzzles, and the convexity of their foreheads.

But if we look for some of those essential changes which would be required to lend even the semblance of a foundation for the theory of Lamarck, respecting the growth of new organs and the gradual obliteration of others, we find nothing of the kind. For in all these varieties of the dog, says Cuvier, the relation of the bones with each other remain essentially the same; the form of the teeth never changes in any perceptible degree, except that in some individuals, one additional false grinder occasionally appears, sometimes on the one side, and sometimes on the other †. The greatest departure from a common type, and it constitutes the maximum of variation as yet known in the animal kingdom, is exemplified in those races of dogs which have a supernumerary toe on the hind foot with the corresponding tarsal bones, a variety analogous to one presented by six-fingered families of the human race ‡.

Lamarck has thrown out as a conjecture, that the wolf may have been the original of the dog, but he has adduced no data to bear out such an hypothesis. " The wolf," observes Dr. Prichard, " and the dog differ, not only with respect to their habits and instincts, which in the brute creation are very uniform within the limits of one species; but some differences have also been pointed out in their internal organization,

* Dureau de la Malle, Ann. des. Sci. Nat. tom. xxi. p. 63. Sept. 1830.
    † Disc. Prel., p. 129, sixth edition.   ‡ Ibid.

particularly in the structure of a part of the intestinal canal*."

It is well known that the horse, the ox, the boar and other domestic animals, which have been introduced into South America, and have run wild in many parts, have entirely lost all marks of domesticity, and have reverted to the original characters of their species. But the dog has also become wild in Cuba, Hayti, and in all the Caribbean islands. In the course of the seventeenth century, they hunted in packs from twelve to fifty, or more in number, and fearlessly attacked herds of wild-boars and other animals. It is natural, therefore, to enquire to what form they reverted ? Now they are said by many travellers to have resembled very nearly the shepherd's dog; but it is certain that they were never turned into wolves. They were extremely savage, and their ravages appear to have been as much dreaded as those of wolves, but when any of their whelps were caught, and brought from the woods to the towns, they grew up in the most perfect submission to man.

As the advocates of the theory of transmutation trust much to the slow and insensible changes which time may work, they are accustomed to lament the absence of accurate descriptions, and figures of particular animals and plants, handed down from the earliest periods of history, such as might have afforded data for comparing the condition of species, at two periods considerably remote. But fortunately, we are in some measure independent of such evidence, for by a singular accident, the priests of Egypt have bequeathed to us, in their cemeteries, that information, which the museums and works of the Greek philosophers have failed to transmit.

For the careful investigation of these documents, we are greatly indebted to the skill and diligence of those naturalists who accompanied the French armies during their brief occupation of Egypt: that conquest of four years, from which we may date the improvement of the modern Egyptians in the arts

* Prichard, Phys. Hist. of Mankind, vol i. p. 96, who cites Professor Güldenstädt.

and sciences, and the rapid progress which has been made of late in our knowledge of the arts and sciences of their remote predecessors. Instead of wasting their whole time as so many preceding travellers had done, in exclusively collecting human mummies, M. Geoffroy and his associates examined diligently, and sent home great numbers of embalmed bodies of consecrated animals, such as the bull, the dog, the cat, the ape, the ichneumon, the crocodile, and the ibis.

To those who have never been accustomed to connect the facts of Natural History with philosophical speculations, who have never raised their conceptions of the end and import of such studies beyond the mere admiration of isolated and beautiful objects, or the exertion of skill in detecting specific differences, it will seem incredible that amidst the din of arms, and the stirring excitement of political movements, so much enthusiasm could have been felt in regard to these precious remains.

In the official report drawn up by the Professors of the Museum at Paris, on the value of these objects, there are some eloquent passages which may appear extravagant, unless we reflect how fully these naturalists could appreciate the bearing of the facts thus brought to light on the past history of the globe.

"It seems," say they, "as if the superstition of the ancient Egyptians had been inspired by Nature, with a view of transmitting to after ages a monument of her history. That extraordinary and whimsical people, by embalming with so much care the brutes which were the objects of their stupid adoration, have left us, in their sacred grottoes, cabinets of zoology almost complete. The climate has conspired with the art of embalming to preserve the bodies from corruption, and we can now assure ourselves by our own eyes what was the state of a great number of species three thousand years ago. We can scarcely restrain the transports of our imagination, on beholding thus preserved with their minutest bones, with the smallest portions of their skin, and in every particular most perfectly recognizable, many an animal, which at Thebes or Memphis, two or three thousand years ago, had its own priests and altars *."

Among the Egyptian mummies thus procured were not only those of numerous wild quadrupeds, birds, and reptiles, but, what was perhaps of still greater importance in deciding the great question under discussion, there were the mummies of domestic animals, among which those above mentioned, the bull, the dog, and the cat, were frequent. Now such was the conformity of the whole of these species to those now living, that there was no more difference, says Cuvier, between them than between the human mummies and the embalmed bodies of men of the present day. Yet some of these animals have since that period been transported by man to almost every variety of climate, and forced to accommodate their habits to new circumstances, as far as their nature would permit. The cat, for example, has been carried over the whole earth, and, within the last three centuries, has been naturalized in every part of the new world, from the cold regions of Canada to the tropical plains of Guiana; yet it has scarcely undergone any perceptible mutation, and is still the same animal which was held sacred by the Egyptians.

Of the ox, undoubtedly there are many very distinct races; but the bull Apis, which was led in solemn processions by the Egyptian priests, did not differ from some of those now living. The black cattle that have run wild in America, where there were many peculiarities in the climate not to be found, perhaps, in any part of the old world, and where scarcely a single plant on which they fed was of precisely the same species, instead of altering their form and habits, have actually reverted to the exact likeness of the aboriginal wild cattle of Europe.

In answer to the arguments drawn from the Egyptian mummies, Lamarck said that they were identical with their living descendants in the same country, because the climate and

* Ann. du Museum, d'Hist. Nat., tom. i. p. 234. 1802. The reporters were MM. Cuvier, Lacépède, and Lamarck.

physical geography of the banks of the Nile have remained unaltered for the last thirty centuries. But why, we ask, have other individuals of these species retained the same characters in so many different quarters of the globe, where the climate and many other conditions are so varied ?

The evidence derived from the Egyptian monuments was not confined to the animal kingdom ; the fruits, seeds, and other portions of twenty different plants, were faithfully preserved in the same manner ; and among these the common wheat was procured by Delille, from closed vessels in the sepulchres of the kings, the grains of which retained not only their form, but even their colour, so effectual has proved the process of embalming with bitumen in a dry and equable climate. No difference could be detected between this wheat and that which now grows in the East and elsewhere, and similar identifications were made in regard to all the other plants.

And here we may observe, that there is an obvious answer to Lamarck's objection *, that the botanist cannot point out a country where the common wheat grows wild, unless in places where it may have been derived from neighbouring cultivation. All naturalists are well aware that the geographical distribution of a great number of species is extremely limited, and that it was to be expected that every useful plant should first be cultivated successfully in the country where it was indigenous, and that, probably, every station which it partially occupied, when growing wild, would be selected by the agriculturist as best suited to it when artificially increased. Palestine has been conjectured, by a late writer on the Cerealia, to have been the original habitation of wheat and barley, a supposition which appears confirmed by Hebrew and Egyptian traditions, and by tracing the migrations of the worship of Ceres, as indicative of the migrations of the plant †.

If we are to infer that some one of the wild grasses has been

* Phil. Zool., tom. i., p. 227.
† L'Origine et la Patrie des Céréales, &c. Ann. des Sci. Nat., tom. ix., p. 61.

transformed into the common wheat, and that some animal of the genus *canis*, still unreclaimed, has been metamorphosed into the dog, merely because we cannot find the domestic dog, or the cultivated wheat, in a state of nature, we may be next called upon to make similar admissions in regard to the camel; for it seems very doubtful whether any race of this species of quadruped is now wild.

But if agriculture, it will be said, does not supply examples of extraordinary changes of form and organization, the horticulturist can, at least, appeal to facts which may confound the preceding train of reasoning. The crab has been transformed into the apple; the sloe into the plum : flowers have changed their colour and become double; and these new characters can be perpetuated by seed,—a bitter plant with wavy sea-green leaves has been taken from the sea-side where it grew like wild charlock, has been transplanted into the garden, lost its saltness, and has been metamorphosed into two distinct vegetables as unlike each other as is each to the parent plant—the red cabbage and the cauliflower. These, and a multitude of analogous facts, are undoubtedly among the wonders of nature, and attest more strongly, perhaps, the extent to which species may be modified, than any examples derived from the animal kingdom. But in these cases we find, that we soon reach certain limits, beyond which we are unable to cause the individuals, descending from the same stock, to vary; while, on the other hand, it is easy to show that these extraordinary varieties could seldom arise, and could never be perpetuated in a wild state for many generations, under any imaginable combination of accidents. They may be regarded as extreme cases brought about by human interference, and not as phenomena which indicate a capability of indefinite modification in the natural world.

The propagation of a plant by buds or grafts, and by cuttings, is obviously a mode which nature does not employ; and this multiplication, as well as that produced by roots and layers, seems merely to operate as an extension of the life of an individual, and not as a reproduction of the species, as happens by seed. All plants increased by the former means retain precisely the peculiar qualities of the individual to which they owe their origin, and, like an individual, they have only a determinate existence ; in some cases longer and in others shorter *. It seems now admitted by horticulturists, that none of our garden varieties of fruit are entitled to be considered strictly permanent, but that they wear out after a time † ; and we are thus compelled to resort again to seeds; in which case, there is so decided a tendency in the seedlings to revert to the original type, that our utmost skill is sometimes baffled in attempting to recover the desired variety.

The different races of cabbages afford, as we have admitted, an astonishing example of deviation from a common type; but we can scarcely conceive them to have originated, much less to have lasted for several generations, without the intervention of man. It is only by strong manures that these varieties have been obtained, and in poorer soils they instantly degenerate. If, therefore, we suppose in a state of nature the seed of the wild Brassica oleracea to have been wafted from the sea-side to some spot enriched by the dung of animals, and to have there become a cauliflower, it would soon diffuse its seed to some comparatively steril soils around, and the offspring would relapse to the likeness of the parent stock, like some individuals which may now be seen growing on the cornice of old London bridge.

But if we go so far as to imagine the soil, in the spot first occupied, to be constantly manured by herds of wild animals, so as to continue as rich as that of a garden, still the variety could not be maintained, because we know that each of these races is prone to fecundate others, and gardeners are compelled to exert the utmost diligence to prevent cross-breeds. The intermixture of the pollen of varieties growing in the poorer soil around, would soon destroy the peculiar characters

---

* Smith's Introduction to Botany, p. 138.   Edit. 1807.
    † See Mr. Knight's Observations, Hort. Trans., vol. ii., p. 160.

of the race which occupied the highly-manured tract; for, if
these accidents so continually happen in spite of us, among
the culinary varieties, it is easy to see how soon this cause
might obliterate every marked singularity in a wild state.

Besides, it is well-known that although the pampered races
which we rear in our gardens for use or ornament, may
often be perpetuated by seed, yet they rarely produce seed
in such abundance, or so prolific in quality, as wild individuals;
so that, if the care of man were withdrawn, the most fertile
variety would always, in the end, prevail over the more steril.

Similar remarks may be applied to the double flowers which
present such strange anomalies to the botanist. The ovarium,
in such cases, is frequently abortive, and the seeds, when prolific, are generally much fewer than where the flowers are
single.

Some curious experiments recently made on the production
of blue instead of red flowers in the Hydrangea hortensis,
illustrate the immediate effect of certain soils on the colours of
the petals. In garden-mould or compost, the flowers are invariably red; in some kinds of bog-earth they are blue; and
the same change is always produced by a particular sort of
yellow loam.

Linnæus was of opinion that the primrose, oxlip, cowslip,
and polyanthus, were only varieties of the same species. The
majority of modern botanists, on the contrary, consider them
to be distinct, although some conceived that the oxlip might
be a cross between the cowslip and the primrose. Mr. Herbert has lately recorded the following experiment:—" I raised
from the natural seed of one umbel of a highly-manured red
cowslip, a primrose, a cowslip, oxlips of the usual and other
colours, a black polyanthus, a hose-in-hose cowslip, and a natural primrose bearing its flower on a polyanthus stalk. From
the seed of that very hose-in-hose cowslip I have since raised a
hose-in-hose primrose. I therefore consider all these to be
only local varieties depending upon soil and situation *." Pro-

* Hort. Trans., vol. iv., p. 19.

fessor Henslow, of Cambridge, has since confirmed this experiment of Mr. Herbert, so that we have an example, not only of the remarkable varieties which the florist can obtain from a common stock, but of the distinctness of analogous races found in a wild state *.

On what particular ingredient, or quality in the earth, these changes depend, has not yet been ascertained †. But gardeners are well aware that particular plants, when placed under the influence of certain circumstances, are changed in various ways according to the species; and as often as the experiments are repeated similar results are obtained. The nature of these results, however, depends upon the species, and they are, therefore, part of the specific character; they exhibit the same phenomena again and again, and indicate certain fixed and invariable relations between the physiological peculiarities of the plant, and the influence of certain external agents. They afford no ground for questioning the instability of species, but rather the contrary; they present us with a class of phenomena which, when they are more thoroughly understood, may afford some of the best tests for identifying species, and proving that the attributes originally conferred, endure so long as any issue of the original stock remains upon the earth.

* Loudon's Mag. of Nat. Hist., Sept. 1830, vol. iii., p. 408.
† Hort. Trans., vol. iii., p. 173.

CHAPTER III.

Variability of a species compared to that of an individual—Species which are susceptible of modification may be altered greatly in a short time, and in a few generations; after which they remain stationary—The animals now subject to man had originally an aptitude to domesticity—Acquired peculiarities which become hereditary have a close connexion with the habits or instincts of the species in a wild state—Some qualities in certain animals have been conferred with a view of their relation to man—Wild elephant domesticated in a few years, but its faculties incapable of further development.

WE endeavoured in the last chapter to show, that a belief in the reality of species is not inconsistent with the idea of a considerable degree of variability in the specific character. This opinion, indeed, is little more than an extension of the idea which we must entertain of the identity of an individual, throughout the changes which it is capable of undergoing.

If a quadruped, inhabiting a cold northern latitude, and covered with a warm coat of hair or wool, be transported to a southern climate, it will often, in the course of a few years, shed a considerable portion of its coat, which it gradually recovers on being again restored to its native country. Even there the same changes are, perhaps, superinduced to a certain extent by the returns of winter and summer. We know that the Alpine hare * and the ermine † become white during winter, and again obtain their full colour during the warmer season; that the plumage of the ptarmigan undergoes a like metamorphosis in colour and quantity, and that the change is equally temporary. We are aware that, if we reclaim some wild animal, and modify its habits and instincts by domestication, it may, if it escapes, become in a few years nearly as wild and untractable as ever; and if the same individual be again retaken, it may be reduced to its former tame state. A plant is placed in a prepared soil in order that the petals of its flowers may multiply, and their colour be heightened or changed; if we

---

* Lepus variabilis.—Pallas.   † Mustela erminea.—Linn.

then withhold our care, the flowers of this same individual become again single. In these, and innumerable other instances, we must suppose that the individual was produced with a certain number of qualities; and, in the case of animals, with a variety of instincts, some of which may or may not be developed according to circumstances, or which, after having been called forth, may again become latent when the exciting causes are removed.

Now the formation of races seems the necessary consequence of such a capability in individuals to vary, if it be a general law that the offspring should very closely resemble the parent. But, before we can infer that there are no limits to the deviation from an original type which may be brought about in the course of an indefinite number of generations, we ought to have some proof that, in each successive generation, individuals may go on acquiring an equal amount of new peculiarities, under the influence of equal changes of circumstances. The balance of evidence, however, inclines most decidedly on the opposite side, for in all cases we find that the quantity of divergence diminishes from the first in a very rapid ratio.

It cannot be objected, that it is out of our power to go on varying the circumstances in the same manner as might happen in the natural course of events during some great geological cycle. For in the first place, where a capacity is given to individuals to adapt themselves to new circumstances, it does not generally require a very long period for its development; if, indeed, such were the case, it is not easy to see how the modification would answer the ends proposed, for all the individuals would die before new qualities, habits, or instincts, were conferred.

When we have succeeded in naturalizing some tropical plant in a temperate climate, nothing prevents us from attempting gradually to extend its distribution to higher latitudes, or to greater elevations above the level of the sea, allowing equal quantities of time, or an equal number of generations for habituating the species to successive increments of cold. But every husbandman and gardener is aware that such experiments will fail; and we are more likely to succeed in making some plants, in the course of the first two generations, support a considerable degree of difference of temperature than a very small difference afterwards, though we persevere for many centuries.

It is the same if we take any other cause instead of temperature; such as the quality of the food, or the kind of dangers to which an animal is exposed, or the soil in which a plant lives. The alteration in habits, form, or organization, is often rapid during a short period; but when the circumstances are made to vary further, though in ever so slight a degree, all modification ceases, and the individual perishes. Thus some herbivorous quadrupeds may be made to feed partially on fish or flesh, but even these can never be taught to live on some herbs which they reject, and which would even poison them, although the same may be very nutritious to other species of the same natural order. So when man uses force or stratagem against wild animals, the persecuted race soon becomes more cautious, watchful, and cunning; new instincts seem often to be developed, and to become hereditary in the first two or three generations; but let the skill and address of man increase, however gradually, no further variation can take place, no new qualities are elicited by the increasing dangers. The alteration of the habits of the species has reached a point beyond which no ulterior modification is possible, however indefinite the lapse of ages during which the new circumstances operate. Extirpation then follows, rather than such a transformation as could alone enable the species to perpetuate itself under the new state of things.

It has been well observed by M. F. Cuvier and M. Dureau de la Malle, that unless some animals had manifested in a wild state an aptitude to second the efforts of man, their domestication would never have been attempted. If they had all resembled the wolf, the fox, and the hyæna, the patience of the experimentalist would have been exhausted by innumerable failures before he at last succeeded in obtaining some imperfect results; so, if the first advantages derived from the cultivation of plants had been elicited by as tedious and costly a process as that by which we now make some slight additional improvement in certain races, we should have remained to this day in ignorance of the greater number of their useful qualities.

It is undoubtedly true, that many new habits and qualities have not only been acquired in recent times by certain races of dogs, but have been transmitted to their offspring. But in these cases it will be observed, that the new peculiarities have an intimate relation to the habits of the animal in a wild state, and therefore do not attest any tendency to departure to an indefinite extent from the original type of the species. A race of dogs employed for hunting deer in the platform of Santa Fé in Mexico, affords a beautiful illustration of a new hereditary instinct. The mode of attack, observes M. Roulin, which they employ, consists in seizing the animal by the belly and overturning it by a sudden effort, taking advantage of the moment when the body of the deer rests only upon the fore-legs. The weight of the animal thus thrown over, is often six times that of its antagonist. The dog of pure breed inherits a disposition to this kind of chase, and never attacks a deer from before while running. Even should the latter, not perceiving him, come directly upon him, the dog steps aside and makes his assault on the flank, whereas other hunting dogs, though of superior strength and general sagacity, which are brought from Europe, are destitute of this instinct. For want of similar precautions, they are often killed by the deer on the spot, the vertebræ of their neck being dislocated by the violence of the shock *.

A new instinct also has become hereditary in a mongrel race of dogs employed by the inhabitants of the banks of the Magdalena, almost excusively in hunting the white-lipped pecari. The address of these dogs consists in restraining their ardour,

* M. Roulin, Ann. des Sci. Nat., tom. xvi. p. 16, 1829.

and attaching themselves to no animal in particular, but keep-
ing the whole herd in check. Now, among these dogs some
are found, which, the very first time they are taken to the
woods, are acquainted with this mode of attack ; whereas, a
dog of another breed starts forward at once, is surrounded
by the pecari, and whatever may be his strength is destroyed
in a moment.

Some of our countrymen, engaged of late in conducting the
principal mining association in Mexico *, carried out with them
some English greyhounds of the best breed, to hunt the hares
which abound in that country.   The great platform which is the
scene of sport is at an elevation of about nine thousand feet
above the level of the sea, and the mercury in the barometer
stands habitually at the height of about nineteen inches.   It
was found that the greyhounds could not support the fatigues
of a long chase in this attenuated atmosphere, and before they
could come up with their prey, they lay down gasping for
breath ; but these same animals have produced whelps which
have grown up, and are not in the least degree incommoded by
the want of density in the air, but run down the hares with as
much ease as the fleetest of their race in this country.

The fixed and deliberate stand of the pointer has with
propriety been regarded as a mere modification of a habit,
which may have been useful to a wild race accustomed to
wind game, and steal upon it by surprise, first pausing for
an instant in order to spring with unerring aim.   The faculty
of the Retriever, however, may justly be regarded as more
inexplicable and less easily referrible to the instinctive passions
of the species.   M. Majendie, says a French writer in a
recently-published memoir, having learnt that there was a race
of dogs in England, which stopped and brought back game of
their own accord, procured a pair, and having obtained a whelp
from them kept it constantly under his eyes, until he had an
opportunity of assuring himself that, without having received any
instruction and on the very first day that it was carried to the

* The Real del Monte Company.

chase, it brought back game with as much steadiness as dogs which had been schooled into the same manœuvre by means of the whip and collar.

Such attainments, as well as the habits and dispositions which the shepherd's dog and many others inherit, seem to be of a nature and extent which we can hardly explain by supposing them to be modifications of instincts necessary for the preservation of the species in a wild state. When such remarkable habits appear in races of this species, we may reasonably conjecture that they were given with no other view than for the use of man and the preservation of the dog which thus obtains protection.

As a general rule, we fully agree with M. F. Cuvier that, in studying the habits of animals, we must attempt, as far as possible, to refer their domestic qualities to modifications of instincts which are implanted in them in a state of nature; and that writer has successfully pointed out, in an admirable essay on the domestication of the mammalia, the true origin of many dispositions which are vulgarly attributed to the influence of education alone*. But we should go too far if we did not admit that some of the qualities of particular animals and plants may have been given solely with a view to the connexion which it was foreseen would exist between them and man—especially when we see that connexion to be in many cases so intimate, that the greater number, and sometimes all the individuals of the species which exist on the earth are in subjection to the human race.

We can perceive in a multitude of animals, especially in some of the parasitic tribes, that certain instincts and organs are conferred for the purpose of defence or attack against some other species. Now if we are reluctant to suppose the existence of similar relations between man and the instincts of many of the inferior animals, we adopt an hypothesis no less violent, though in the opposite extreme to that which has led some to imagine the whole animate and inanimate creation to have been

* Mem. du Mus. d'Hist. Nat.—Jameson, Ed. New Phil. Journ., Nos. 6, 7, 8.

made solely for the support, gratification, and instruction of mankind.

Many species most hostile to our persons or property multiply in spite of our efforts to repress them ; others, on the contrary, are intentionally augmented many hundred-fold in number by our exertions. In such instances we must imagine the relative resources of man and of species, friendly or inimical to him, to have been prospectively calculated and adjusted. To withhold assent to this supposition would be to refuse what we must grant in respect to the economy of Nature in every other part of the organic creation ; for the various species of contemporary plants and animals have obviously their relative forces nicely balanced, and their respective tastes, passions, and instincts, so contrived, that they are all in perfect harmony with each other. In no other manner could it happen, that each species surrounded as it is by countless dangers should be enabled to maintain its ground for periods of considerable duration.

The docility of the individuals of some of our domestic species extending, as it does, to attainments foreign to their natural habits and faculties, may perhaps have been conferred with a view to their association with man. But lest species should be thereby made to vary indefinitely, we find that such habits are never transmissible by generation.

A pig has been trained to hunt and point game with great activity and steadiness * ; and other learned individuals, of the same species, have been taught to spell ; but such fortuitous acquirements never become hereditary, for they have no relation whatever to the exigencies of the animal in a wild state, and cannot therefore be developments of any instinctive propensities.

An animal in domesticity, says M. F. Cuvier, is not essentially in a different situation in regard to the feeling of restraint from one left to itself. It lives in society without

* In the New Forest, near Ringwood, Hants, by Mr. Toomer, keeper of Broomy Lodge.

constraint, because without doubt it was a social animal, and it conforms itself to the will of man, because it had a chief to which in a wild state it would have yielded obedience.   There is nothing in its new situation that is not conformable to its propensities ; it is satisfying its wants by submission to a master, and makes no sacrifice of its natural inclinations.   All the social animals when left to themselves form herds more or less numerous, and all the individuals of the same herd know each other, are mutually attached, and will not allow a strange individual to join them.   In a wild state, moreover, they obey some individual, which by its superiority has become the chief of the herd.   Our domestic species had originally this sociability of disposition, and no solitary species, however easy it may be *to tame it*, has yet afforded true domestic races.   We merely, therefore, develope to our own advantage, propensities which propel the individuals of certain species to draw near to their fellows.

The sheep which we have reared is induced to follow us, as it would be led to follow the flock among which it was brought up ; and when individuals of gregarious species have been accustomed to one master, it is he alone whom they acknowledge as their chief, he only whom they obey.— " The elephant only allows himself to be led by the carnac whom he has adopted ; the dog itself, reared in solitude with its master, manifests a hostile disposition towards all others ; and everybody knows how dangerous it is to be in the midst of a herd of cows, in pasturages that are little frequented, when they have not at their head the keeper who takes care of them."

" Everything, therefore, tends to convince us, that formerly men were only, with regard to the domestic animals, what those who are particularly charged with the care of them still are, namely, members of the society, which these animals form among themselves, and that they are only distinguished in the general mass by the authority which they have been enabled to assume from their superiority of intellect.   Thus, every social animal which recognizes man as a member, and as the chief of its

herd, is a domestic animal.   It might even be said that from
the moment when such an animal admits man as a member of
its society, it is domesticated, as man could not enter into such
a society without becoming the chief of it *."

But the ingenious author whose observations we have here
cited, admits that the obedience which the individuals of many
domestic species yield indifferently to every person is without
analogy in any state of things which could exist previously
to their subjugation by man.   Each troop of wild horses, it is
true, has some stallion for its chief, who draws after him all
the individuals of which the herd is composed ; but when a
domesticated horse has passed from hand to hand, and has
served several masters, he becomes equally docile towards *any*
person, and is subjected to the whole human race.   It seems
fair to presume, that the capability in the instinct of the horse
to be thus modified, was given to enable the species to render
greater services to man ; and, perhaps, the facility with which
many other acquired characters become hereditary in various
races of the horse, may be explicable only on a like supposition.
The amble, for example, a pace to which the domestic races
in Spanish America are exclusively trained, has, in the course
of several generations, become hereditary, and is assumed by
all the young colts before they are broken in *.

It seems also reasonable to conclude, that the power bestowed on the horse, the dog, the ox, the sheep, the cat, and
many species of domestic fowls, of supporting almost every
climate, was given expressly to enable them to follow man
throughout all parts of the globe—in order that we might
obtain their services, and they our protection.   If it be objected
that the elephant, which, by the union of strength, intelligence,
and docility, can render the greatest services to mankind, is
incapable of living in any but the warmest latitudes, we may
observe, that the quantity of vegetable food required by this

* Mem. du Mus. d' Hist. Nat.
† Dureau de la Malle, Ann. des Sci. Nat., tom. xxi. p. 58.

quadruped would render its maintenance, in the temperate zone, too costly, and in the arctic impossible.

Among the changes superinduced by man, none appear, at first sight, more remarkable than the perfect tameness of certain domestic races. It is well known, that at however early an age we obtain possession of the young of many unreclaimed races, they will retain, throughout life, a considerable timidity and apprehensiveness of danger; whereas, after one or two generations, the descendants of the same will habitually place the most implicit confidence in man. There is good reason, however, to suspect that such changes are not without analogy in a state of nature, or, to speak more correctly, in situations where man has not interfered.

Thus Dr. Richardson informs us, in his able history of the habits of North American animals, that " in the retired parts of the mountains, where the hunters had seldom penetrated, there is no difficulty in approaching the Rocky Mountain sheep, which there exhibit *the simplicity of character so remarkable in the domestic species;* but where they have been often fired at, they are exceedingly wild, alarm their companions, on the approach of danger, by a hissing noise, and scale the rocks with a speed and agility that baffles pursuit *."

It is probable, therefore, that as man, in diffusing himself over the globe, has tamed many wild races, so also he has made many tame races wild. Had some of the larger carnivorous beasts, capable of scaling the rocks, máde their way into the North American mountains before our hunters, a similar alteration in the instincts of the sheep would doubtless have been brought about.

No animal affords a more striking illustration of the principal points we have been endeavouring to establish than the elephant. For in the first place, the wonderful sagacity with which he accommodates himself to the society of man, and the new habits which he contracts are not the result of time nor of modifications produced in the course of many generations.

* Fauna Boreali-Americana, page 273.

These animals will breed in captivity, as is now ascertained in opposition to the vulgar opinion of many modern naturalists, and in conformity to that of the ancients Ælian and Columella*. Yet it has always been the custom, as the least expensive mode of obtaining them, to capture wild individuals in the forests, usually when full grown, and in a few years after they are taken, sometimes, it is said, in the space of a few months, their education is completed.

Had the whole species been domesticated from an early period in the history of man, like the camel, their superior intelligence would doubtless have been attributed to their long and familiar intercourse with the lord of the creation : but we know that a few years is sufficient to bring about this wonderful change of habits ; and, although the same individual may continue to receive tuition for a century afterwards, yet it makes no further progress in the general development of its faculties. Were it otherwise, indeed, the animal would soon deserve more than the poet's epithet of " half-reasoning."

From the authority of our countrymen employed in the late Burmese war, it appears, in corroboration of older accounts, that when elephants are required to execute extraordinary tasks, they may be made to understand that they will receive unusual rewards. Some favourite dainty is shown to them, in the hope of acquiring which, the work is done. And so perfectly does the nature of the contract appear to be understood, that the breach of it, on the part of the master, is often attended with danger. In this case, a power has been given to the species to adapt their social instincts to new circumstances with surprising rapidity ; but the extent of this change is defined by strict and arbitrary limits. There is no indication of a tendency to continued divergence from certain attributes with which the elephant was originally endued, no ground whatever for anticipating, that in thousands of centuries any material alteration could ever be effected. All that

* Mr. Corse on the Habits, &c. of the Elephant, Phil. Trans. 1799.

we can infer from analogy is, that some useful and peculiar races might probably be formed, if the experiment were fairly tried, and that some individual characteristic, now only casual and temporary, might be perpetuated by generation.

In all cases, therefore, where the domestic qualities exist in animals, they seem to require no lengthened process for their development, and they appear to have been wholly denied to some classes, which from their strength and social nature might have rendered great services to man; as, for example, the greater part of the quadrumana. The orang-outang, indeed, which for its resemblance in form to man, and apparently for no other good reason, has been assumed, by Lamarck, to be the most perfect of the inferior animals, has been tamed by the savages of Borneo, and made to climb lofty trees, and to bring down the fruit. But he is said to yield to his masters an unwilling obedience, and to be held in subjection only by severe discipline. We know nothing of the faculties of this animal which can suggest the idea that it rivals the elephant in intelligence, much less anything which can countenance the dreams of those who have fancied that it might have been transmuted into " the dominant race." One of the baboons of Sumatra (*Simia carpolegus*) appears to be more docile, and is frequently trained by the inhabitants to ascend trees for the purpose of gathering cocoa-nuts, a service in which the animal is very expert. He selects, says Sir Stamford Raffles, the ripe nuts with great judgment, and pulls no more than he is ordered *. The capuchin and cacajao monkeys are, according to Humboldt, taught to ascend trees in the same manner, and to throw down fruit on the banks of the lower Orinoco †.

We leave it to the Lamarckians to explain, how it happens that those same savages of Borneo have not themselves acquired, by dint of longing for many generations for the power of climbing trees, the elongated arms of the orang, or even the prehen-

---

* Linn. Trans. vol. xiii. p. 244.

† Pers. Narr. of Travels to the Equinoctial Regions of the New Continent, in the years 1799-1804.

sile tails of some American monkeys. Instead of being reduced to the necessity of subjugating stubborn and untractable brutes, we should naturally have anticipated " that their wants would have excited them to efforts, and that continued efforts would have given rise to new organs ;" or, rather, to the re-acquisition of organs which, in a manner irreconcileable with the principle of the *progressive* system, have grown obsolete in tribes of men which have such constant need of them.

It follows, then, from the different facts which we have considered in this chapter, that a short period of time is generally sufficient to effect nearly the whole change which an alteration of external circumstances can bring about in the habits of a species, and that such capacity of accommodation to new circumstances is enjoyed in very different degrees by different species.

Certain qualities appear to be bestowed exclusively with a view to the relations which are destined to exist between different species, and, among others, between certain species and man ; but these latter are always so nearly connected with the original habits and propensities of each species in a wild state, that they imply no indefinite capacity of varying from the original type. The acquired habits, derived from human tuition, are rarely transmitted to the offspring ; and when this happens, it is almost universally the case with those merely which have some obvious connexion with the attributes of the species when in a state of independence.

CHAPTER IV.

WE have yet to consider another class of phenomena, those relating to the production of hybrids, which have been regarded in a very different light with reference to their bearing on the question of the permanent distinctness of species ; some natu- ralists considering them as affording the strongest of all proofs in favour of the reality of species ; others, on the contrary, appealing to them as countenancing the opposite doctrine, that all the varieties of organization and instinct now exhibited in the animal and vegetable kingdoms, may have been propagated from a small number of original types.

In regard to the mammifers and birds, it is found that no sexual union will take place between races which are remote from each other in their habits and organization ; and it is only in species that are very nearly allied that such unions produce offspring. It may be laid down as a general rule, admitting of very few exceptions among quadrupeds, that the hybrid progeny is steril, and there seem to be no well-authenticated examples of the continuance of the mule race beyond one gene- ration. The principal number of observations and experiments relate to the mixed offspring of the horse and the ass ; and in this case it is well established, that the male-mule can generate and the female-mule produce. Such cases occur in Spain and

E

Italy, and much more frequently in the West Indies and New Holland; but these mules have never bred in cold climates, seldom in warm regions, and still more rarely in temperate countries.

The hybrid offspring of the female-ass and the stallion, the γιννος of Aristotle, and the hinnus of Pliny, differs from the mule, or the offspring of the ass and mare. In both cases, says Buffon, these animals retain more of the mother than of the father, not only in the magnitude but in the figure of the body; whereas, in the form of the head, limbs, and tail, they bear a greater resemblance to the father. The same naturalist infers, from various experiments respecting cross-breeds between the he-goat and ewe, the dog and she-wolf, the goldfinch and canary-bird, that the male transmits his sex to the greatest number, and that the preponderance of males over females exceeds that which prevails where the parents are of the same species.

The celebrated John Hunter has observed, that the true distinction of species must ultimately be gathered from their incapacity of propagating with each other, and producing offspring capable of again continuing itself. He was unwilling, however, to admit, that the horse and the ass were of the same species, because some rare instances had been adduced of the breeding of mules, which he attributed to a degree of monstrosity in the organs of the mule, for these he suggested might not have been those of a mixed animal, but those of the mare or female-ass. " This, he argues, is not a far-fetched idea, for true species produce monsters, and many animals of distinct sex are incapable of breeding at all; and as we find nature, in its greatest perfection, deviating from general principles, why may it not happen likewise in the production of mules, so that sometimes a mule shall breed from the circumstance of its being a monster respecting mules?"

Yet, in the same memoir, this great anatomist inferred that the wolf, the dog, and the jackal, were all of one species, because he had found, by two experiments, that the dog would breed, both with the wolf and the jackal; and that the mule, in each case, would breed again with the dog. In these cases, however, we may observe, that there was always one parent at least of pure breed, and no proof was obtained that a true hybrid race could be perpetuated; a fact of which we believe no examples are yet recorded, either in regard to mixtures of the horse and ass, or any other of the mammalia.

Should the fact be hereafter ascertained, that two mules can propagate their kind, we must still inquire whether the offspring may not be regarded in the light of a monstrous birth, proceeding from some accidental cause, or rather, to speak more philosophically, from some general law not yet understood, but which may not be permitted permanently to interfere with those laws of generation, whereby species may, in general, be prevented from becoming blended. If, for example, we discovered that the progeny of a mule race degenerated greatly in the first generation, in force, sagacity, or any attribute necessary for its preservation in a state of nature, we might infer that, like a monster, it is a mere temporary and fortuitous variety. Nor does it seem probable that the greater number of such monsters could ever occur unless obtained by art; for in Hunter's experiments, stratagem or force was, in most instances, employed to bring about the irregular connexion *.

It seems rarely to happen that the mule offspring is truly intermediate in character between the two parents. Thus Hunter mentions, that, in his experiments, one of the hybrid pups resembled the wolf much more than the rest of the litter; and we are informed by Wiegmann, that in a litter lately obtained in the Royal Menagerie at Berlin, from a white pointer and a she-wolf, two of the cubs resembled the common wolf-dog, but the third was like a pointer with hanging ears.

There is, undoubtedly, a very close analogy between these phenomena and those presented by the intermixture of distinct races of the same species, both in the inferior animals and in

---

* Phil. Trans. 1787. Additional Remarks, Phil. Trans. 1789.

man.   Dr. Prichard, in his " Physical History of Mankind," cites examples where the peculiarities of the parents have been transmitted very unequally to the offspring ; as where children, entirely white, or perfectly black, have sprung from the union of the European and the negro.   Sometimes the colour, or other peculiarities of one parent, after having failed to show themselves in the immediate progeny, reappear in a subsequent generation, as where a white child is born of two black parents, the grandfather having been a white *.

The same author judiciously observes, that if different species mixed their breed, and hybrid races were often propagated, the animal world would soon present a scene of confusion ; its tribes would be everywhere blended together, and we should, perhaps, find more hybrid creatures than genuine and uncorrupted races †.

The history of the vegetable kingdom has been thought to afford more decisive evidence in favour of the theory of the formation of new and permanent species from hybrid stocks.   The first accurate experiments in illustration of this curious subject appear to have been made by Kölreuter, who obtained a hybrid from two species of Tobacco, *Nicotiana rustica* and *N. paniculata,* which differ greatly in the shape of their leaves, the colour of the corolla, and the height of the stem.   The stigma of a female plant of *N. rustica* was impregnated with the pollen of a male plant of *N. paniculata.*   The seed ripened and produced a hybrid which was intermediate between the two parents, and which, like all the hybrids which this botanist brought up, had imperfect stamens.   He afterwards impregnated this hybrid with the pollen of *N. paniculata,* and obtained plants which much more resembled the last.   This he continued through several generations, until, by due perseverance, he actually changed the *Nicotiana rustica* into the *Nicotiana paniculata.*

The plan of impregnation adopted, was the cutting off of the anthers of the plant intended for fructification before they had

* Vol. i., p. 217.   † Ibid., vol. i., p. 97.

shed pollen, and then laying on foreign pollen upon the stigma. The same experiment has since been repeated, with success, by Wiegmann, who found that he could bring back the hybrids to the exact likeness of either parent, by crossing them a sufficient number of times.

The blending of the characters of the parent stocks, in many other of Weigmann's experiments, was complete ; the colour and shape of the leaves and flowers, and even the scent, being intermediate, as in the offspring of the two species of verbascum. An intermarriage, also, between the common onion and the leek (*Allium cepa* and *A. porrum*) gave a mule plant, which, in the character of its leaves and flowers, approached most nearly to the garden onion, but had the elongated bulbous root and smell of the leek.

The same botanist remarks, that vegetable hybrids, when not strictly intermediate, more frequently approach the female than the male parent species, *but they never exhibit characters foreign to both.* A re-cross with one of the original stocks, generally causes the mule plant to revert towards that stock ; but this is not always the case, the offspring sometimes continuing to exhibit the character of a full hybrid.

In general, the success attending the production and perpetuity of hybrids among plants, depends, as in the animal kingdom, on the degree of proximity between the species intermarried. If their organization be very remote, impregnation never takes place ; if somewhat less distant, seeds are formed, but always imperfect and steril. The next degree of relationship yields hybrid seedlings, but these are barren ; and it is only when the parent species are very nearly allied, that the hybrid race may be perpetuated for several generations. Even in this case the best authenticated examples seem confined to the crossing of hybrids with individuals of pure breed. In none of the experiments most accurately detailed does it appear that both the parents were mules.

Wiegmann diversified, as much as possible, his mode of bringing about these irregular unions among plants. He often sowed parallel rows, near to each other, of the species from which he desired to breed, and instead of mutilating, after Kölreuter's fashion, the plants of one of the parent stocks, he merely washed the pollen off their anthers. The branches of the plants, in each row, were then gently bent towards each other and intertwined, so that the wind, and numerous insects as they passed from the flowers of one to those of the other species, carried the pollen and produced fecundation.

The same observer saw a good exemplification of the manner in which hybrids may be formed in a state of nature. Some wallflowers and pinks had been growing in a garden, in a dry sunny situation, and their stigmas had been ripened so as to be moist, and to absorb pollen with avidity, although their anthers were not yet developed. These stigmas became impregnated by pollen, blown from some other adjacent plants of the same species, but had they been of different species, and not too remote in their organization, mule races must have resulted.

When, indeed, we consider how busily some insects have been shown to be engaged in conveying anther-dust from flower to flower, especially bees, flower-eating beetles, and the like, it seems a most enigmatical problem how it can happen, that promiscuous alliances between distinct species are not perpetually occurring.

How continually do we observe the bees diligently employed in collecting the red and yellow powder by which the stamens of flowers are covered, loading it on their hind legs, and carrying it to their hive for the purpose of feeding their young! In thus providing for their own progeny, these insects assist materially the process of fructification*. Few of our readers need be reminded, that the stamens in certain plants grow on different blossoms from the pistils, and unless the summit of the pistil be touched with the fertilizing dust, the fruit does not swell, nor the seed arrive at maturity. It is by the help of bees chiefly, that the development of the fruit of many such species

---

* See Barton on the Geography of Plants, p. 67.

is secured, the powder which they have collected from the stamens being unconsciously left by them in visiting the pistils.

How often, during the heat of a summer's day, do we see the males of diœcious plants, such as the yew-tree, standing separate from the females, and sending off into the air, upon the slightest breath of wind, clouds of buoyant pollen! That the zephyr should so rarely intervene to fecundate the plants of one species with the anther-dust of others, seems almost to realize the converse of the miracle believed by the credulous herdsmen of the Lusitanian mares—

> Ore omnes versæ in Zephyrum, stant rupibus altis,
> Exceptantque leves auras : et sæpe sine ullis
> Conjugiis, vento gravidæ, mirabile dictu *.

But, in the first place, it appears that there is a natural aversion in plants, as well as in animals, to irregular sexual unions ; and in most of the successful experiments in the animal and vegetable world, some violence has been used, in order to procure impregnation. The stigma imbibes, slowly and reluctantly, the granules of the pollen of another species, even when it is abundantly covered with it ; and if it happen that, during this period, ever so slight a quantity of the anther-dust of its own species alight upon it, this is instantly absorbed, and the effect of the foreign pollen destroyed. Besides, it does not often happen that the male and female organs of fructification, in different species, arrive at a state of maturity at precisely the same time. Even where such synchronism does prevail, so that a cross impregnation is effected, the chances are very numerous against the establishment of a hybrid race.

If we consider the vegetable kingdom generally, it must be recollected, that even of the seeds which are well ripened, the greater part are either eaten by insects, birds, and other animals, or decay for want of room and opportunity to germinate. Unhealthy plants are the first which are cut off by causes prejudicial to the species, being usually stifled by more vigorous individuals of their own kind. If, therefore, the relative fecundity or

---

* Georg. lib. iii. 273.

hardiness of hybrids be in the least degree inferior, they cannot maintain their footing for many generations, even if they were ever produced beyond one generation in a wild state. In the universal struggle for existence, the right of the strongest eventually prevails ; and the strength and durability of a race depends mainly on its prolificness, in which hybrids are acknowledged to be deficient.

Centaurea hybrida, a plant which never bears seed, and is supposed to be produced by the frequent intermixture of two well-known species of Centaurea, grows wild upon a hill near Turin. *Ranunculus lacerus,* also steril, has been produced accidentally at Grenoble, and near Paris, by the union of two Ranunculi ; but this occurred in gardens *.

Mr. Herbert, in one of his ingenious papers on mule plants, endeavours to account for their non-occurrence in a state of nature, from the circumstance that all the combinations that were likely to occur, have already been made many centuries ago, and have formed the various species of botanists ; but in our gardens, he says, whenever species, having a certain degree of affinity to each other, are transported from different countries, and brought for the first time into contact, they give rise to hybrid species †. But we have no data, as yet, to warrant the conclusion, that a single permanent hybrid race has ever been formed, even in gardens, by the intermarriage of two allied species brought from distant habitations. Until some fact of this kind is fairly established, and a new species, capable of perpetuating itself in a state of perfect independence of man, can be pointed out, we think it reasonable to call in question entirely this hypothetical source of new species. That varieties do sometimes spring up from cross breeds, in a natural way, can hardly be doubted, but they probably die out even more rapidly than races propagated by grafts or layers.

Decandolle, whose opinion on a philosophical question of this kind deserves the greatest attention, has observed, in his Essay

---

* Hon. and Rev. W. Herbert, Hort. Trans., vol. iv., p. 41.   † Ibid.

on Botanical Geography, that the *varieties* of plants range themselves under two general heads : those produced by external circumstances, and those formed by hybridity. After adducing various arguments to show that neither of these causes can explain the permanent diversity of plants indigenous in different regions, he says, in regard to the crossing of races, " I can perfectly comprehend, without altogether sharing the opinion, that where many species of the same genera occur near together, hybrid species may be formed, and I am aware that the great number of species of certain genera which are found in particular regions, may be explained in this manner ; but I am unable to conceive how any one can regard the same explanation as applicable to species which live naturally at great distances.   If the three larches, for example, now known in the world, lived in the same localities, I might then believe that one of them was the produce of the crossing of the two others; but I never could admit that the Siberian species has been produced by the crossing of those of Europe and America.   I see, then, that there exist, in organized beings, permanent differences which cannot be referred to any one of the actual causes of variation, and these differences are what constitute *species* *."

The most decisive arguments, perhaps, amongst many others, against the probability of the derivation of permanent species from cross breeds, are to be drawn from the fact alluded to by Decandolle, of species having a close affinity to each other occurring in distinct botanical provinces, or countries inhabited by groups of distinct species of indigenous plants. For in this case naturalists, who are not prepared to go the whole length of the transmutationists, are under the necessity of admitting, that in some cases species which approach very near to each other in their characters, were so created from their origin ; an admission fatal to the idea of its being a general law of nature, that a few original types only should be formed,

---

* Essai Elémentaire, &c. 3me. partie.

and that all intermediate races should spring from the intermixture of those stocks.

This notion, indeed, is wholly at variance with all that we know of hybrid generation ; for the phenomena entitle us to affirm, that had the types been at first somewhat distant, *no cross-breeds would ever have been produced*, much less those prolific races which we now recognise as distinct species.

In regard, moreover, to the permanent propagation of hybrid races among animals, insuperable difficulties present themselves, when we endeavour to conceive the blending together of the different instincts and propensities of two species, so as to insure the preservation of the intermediate race. The common mule, when obtained by human art, may be protected by the power of man ; but in a wild state, it would neither have precisely the same wants as the horse or the ass : and if, in consequence of some difference of this kind, it strayed from the herd, it would soon be hunted down by beasts of prey and destroyed.

If we take some genus of insects, such as the bee, we find that each of the numerous species has some difference in its habits, its mode of collecting honey, or constructing its dwelling, or providing for its young, and other particulars. In the case of the common hive-bee, the workers are described, by Kirby and Spence, as being endowed with no less than thirty distinct instincts *. So also we find that amongst a most numerous class of spiders, there are nearly as many different modes of spinning their webs as there are species. When we recollect how complicated are the relations of these instincts with co-existing species, both of the animal and vegetable kingdoms, it is scarcely possible to imagine that a bastard race could spring from the union of two of these species, and retain just so much of the qualities of each parent-stock as to preserve its ground in spite of the dangers which surround it.

We should also ask, if a few generic types alone have been *created* among insects, and the intermediate species have pro-

* Intr. to Entom., vol. ii., p. 504. Ed. 1817.

ceeded from hybridity, where are those original types, com-
bining, as they ought to do, the elements of all the instincts
which have made their appearance in the numerous derivative
races? So also in regard to animals of all classes, and of plants;
if species in general are of hybrid origin, where are the stocks
which combine in themselves the habits, properties, and organs,
of which all the intervening species ought to afford us mere
modifications?

We shall now conclude this subject by summing up, in a
few words, the results to which the consideration of the phenomena of hybrids has led us. It appears that the aversion of
individuals of distinct species to the sexual union is common
to animals and plants, and that it is only when the species
approach near to each other, in their organization and habits,
that any offspring are produced from their connexion. Mules
are of extremely rare occurrence in a state of nature, and no
examples are yet known of their having procreated in a wild
state. But it has been proved, that hybrids are not universally steril, provided the parent stocks have a near affinity to
each other, although the continuation of the mixed race, for
several generations, appears hitherto to have been obtained
only by crossing the hybrids with individuals of pure species,
an experiment which by no means bears out the hypothesis
that a true hybrid race could ever be permanently established.

Hence we may infer, that aversion to sexual intercourse is,
in general, a good test of the distinctness of original stocks, or
of *species*, and the procreation of hybrids is a proof of the very
near affinity of species. Perhaps, hereafter, the number of
generations for which hybrids may be continued, before the
race dies out (for it seems usually to degenerate rapidly), may
afford the zoologist and botanist an experimental test of the
difference in the degree of affinity of allied species.

We may also remark, that if it could have been shown that
a single permanent species had ever been produced by hybridity (of which there is no satisfactory proof), it might certainly have lent some countenance to the notions of the ancients

respecting the gradual deterioration of created things, but none whatever to Lamarck's theory of their progressive perfectibility; for observations have hitherto shown that there is a tendency, in mule animals and plants, to degenerate in organization.

We have already remarked, that the theory of progressive development arose from an attempt to ingraft the doctrines of the transmutationists upon one of the most popular generalizations in geology. But modern geological researches have almost destroyed every appearance of that gradation in the successive groups of animate beings, which was supposed to indicate the slow progress of the organic world from the more simple to the more compound structure. In the more modern formations, we find clear indications that the highest orders of the terrestrial mammalia were fully represented during several successive epochs; but, in the monuments which we have hitherto examined of more remote eras, in which there are as yet discovered few fluviatile, and perhaps no lacustrine formations, and, therefore, scarcely any means of obtaining an insight into the zoology of the then existing continents, we have only as yet found one example of a mammiferous quadruped. The recent origin of man, and the absence of all signs of any rational being holding an analogous relation to former states of the animate world, affords one, and the only reasonable argument, in support of the hypothesis of a progressive scheme, but none whatever in favour of the fancied evolution of one species out of another.

When the celebrated anatomist, Camper, first attempted to estimate the degrees of sagacity of different animals, and of the races of man, by the measurement of the facial angle, some speculators were bold enough to affirm, that certain simiæ differed as little from the more savage races of men, as do these from the human race in general; and that a scale might be traced from " apes with foreheads villanous low," to the African variety of the human species, and from that to the European. The facial angle was measured by drawing a line from the prominent centre of the forehead to the most advanced part of the lower jaw-bone, and observing the angle which it made with the horizontal line ; and it was affirmed, that there was a regular series from birds to the mammalia.

The gradation from the dog to the monkey was said to be perfect, and from that again to man. One of the ape tribe has a facial angle of 42°, and another, which approximated nearest to man in figure, an angle of 50°. To this succeeds (longo sed proximus intervallo) the head of the African negro, which, as well as that of the Kalmuc, forms an angle of 70°, while that of the European contains 80°. The Roman painters preferred the angle of 95°, and the character of beauty and sublimity, so striking in some works of Grecian sculpture, as in the head of Apollo, and in the Medusa of Sisocles, is given by an angle which amounts to 100° *.

A great number of valuable facts and curious analogies in comparative anatomy, were brought to light during the investigations which were made by Camper, John Hunter, and others, to illustrate this scale of organization ; and their facts and generalizations must not be confounded with the fanciful systems which White and others deduced from them †.

That there is some connexion between an elevated and capacious forehead in certain races of men, and a large development of the intellectual faculties, seems highly probable ; and that a low facial angle is frequently accompanied with inferiority of mental powers, is certain ; but the attempt to trace a graduated scale of intelligence through the different species of animals accompanying the modifications of the form of the skull, is a mere visionary speculation. It has been found necessary to exaggerate the sagacity of the ape tribe at the expense of the dog, and strange contradictions have arisen in the conclusions deduced from the structure of the elephant, some anatomists being disposed to deny the quadruped the intelligence which he really possesses, because they found that the volume of his brain was small in comparison to that of the other mammalia, while others were inclined to magnify extravagantly

---

* Prichard, Phys. Hist. of Mankind, vol. i., p. 159.

† Ch. White on the regular Gradation in Man, &c., 1799.

the superiority of its intellect, because the vertical height of its skull is so great when compared to its horizontal length.

It would be irrelevant to our subject if we were to enter into a farther discussion on these topics, because, even if a graduated scale of organization and intelligence could have been established, it would prove nothing in favour of a tendency, in each species, to attain a higher state of perfection. We may refer the reader to the writings of Blumenbach, Prichard, Lawrence, and others, for convincing proofs that the varieties of form, colour, and organization of different races of men, are perfectly consistent with the generally received opinion, that all the individuals of the species have originated from a single pair ; and while they exhibit in man as many diversities of a physiological nature, as appear in any other species, they confirm also the opinion of the slight deviation from a common standard of which a species is capable.

The power of existing and multiplying in every latitude, and in every variety of situation and climate, which has enabled the great human family to extend itself over the habitable globe, is partly, says Lawrence, the result of physical constitution, and partly of the mental prerogative of man. If he did not possess the most enduring and flexible corporeal frame, his arts would not enable him to be the inhabitant of all climates, and to brave the extremes of heat and cold, and the other destructive influences of local situation *. Yet, notwithstanding this flexibility of bodily frame, we find no signs of indefinite departure from a common standard, and the intermarriages of individuals of the most remote varieties are not less fruitful than between those of the same tribe.

There is yet another department of anatomical discovery, to which we must not omit some allusion, because it has appeared to some persons to afford a distant analogy, at least, to that progressive development by which some of the inferior species may have been gradually perfected into those of more complex organization. Tieddemann found, and his discoveries have been

* Lawrence, Lectures on Phys. Zool. and Nat. Hist. of Man, p. 192. Ed. 1823.

most fully confirmed and elucidated by M. Serres, that the brain of the fœtus, in the highest class of vertebrated animals, assumes, in succession, the various forms which belong to fishes, reptiles, and birds, before it acquires those additions and modifications which are peculiar to the mammiferous tribe. So that in the passage from the embryo to the perfect mammifer, there is a typical representation, as it were, of all those transformations which the primitive species are supposed to have undergone, during a long series of generations, between the present period and the remotest geological era.

If you examine the brain of the mammalia, says M. Serres, at an early stage of uterine life, you perceive the cerebral hemispheres consolidated, as in fish, in two vesicles isolated one from the other; at a later period, you see them affect the configuration of the cerebral hemispheres of reptiles; still later again, they present you with the forms of those of birds; finally, they acquire, at the era of birth, and sometimes later, the permanent forms which the adult mammalia present.

The cerebral hemispheres, then, only arrive at the state which we observe in the higher animals by a series of successive metamorphoses. If we reduce the whole of these evolutions to four periods, we shall see that in the first are born the cerebral lobes of fishes, and this takes place homogeneously in all classes. The second period will give us the organization of reptiles; the third the brain of birds; and the fourth the complex hemispheres of mammalia.

If we could develop the different parts of the brain of the inferior classes, we should make in succession a reptile out of a fish, a bird out of a reptile, and a mammiferous quadruped out of a bird. If, on the contrary, we could starve this organ in the mammalia, we might reduce it successively to the condition of the brain of the three inferior classes.

Nature often presents us with this last phenomenon in monsters, but never exhibits the first. Among the various deformities which organized beings may experience, they never pass the limits of their own classes to put on the forms of the class above them. Never does a fish elevate itself so as to assume the form of the brain of a reptile; nor does the latter ever attain that of birds; nor the bird that of the mammifer. It may happen that a monster may have two heads, but the conformation of the brain always remains circumscribed narrowly within the limits of its class *.

It will be observed, that these curious phenomena disclose, in a highly interesting manner, the unity of plan that runs through the organization of the whole series of vertebrated animals; but they lend no support whatever to the notion of a gradual transmutation of one species into another, least of all of the passage, in the course of many generations, from an animal of a more simple, to one of a more complex structure. On the contrary, were it not for the sterility imposed on monsters, as well as on hybrids in general, the argument to be derived from Tieddemann's discovery, like that deducible from experiments respecting hybridity, would be in favour of the successive *degeneracy*, rather than the perfectibility, in the course of ages, of certain classes of organic beings.

For the reasons, therefore, detailed in this and the two preceding chapters, we draw the following inferences, in regard to the reality of *species* in nature.

First, That there is a capacity in all species to accommodate themselves, to a certain extent, to a change of external circumstances, this extent varying greatly according to the species.

2dly. When the cnange of situation which they can endure is great, it is usually attended by some modifications of the form, colour, size, structure, or other particulars ; but the mutations thus superinduced are governed by constant laws, and the capability of so varying forms part of the permanent specific character.

3dly. Some acquired peculiarities of form, structure, and instinct, are transmissible to the offspring; but these consist

---

* E. R. A. Serres, Anatomie Comparée du Cerveau, illustrated by numerous plates, tom. i., 1824.

of such qualities and attributes only as are intimately related to the natural wants and propensities of the species.

4thly. The entire variation from the original type, which any given kind of change can produce, may usually be effected in a brief period of time, after which no farther deviation can be obtained by continuing to alter the circumstances, though ever so gradually,—indefinite divergence, either in the way of improvement or deterioration, being prevented, and the least possible excess beyond the defined limits being fatal to the existence of the individual.

5thly. The intermixture of distinct species is guarded against by the aversion of the individuals composing them to sexual union, or by the sterility of the mule offspring. It does not appear that true hybrid races have ever been perpetuated for several generations, even by the assistance of man; for the cases usually cited relate to the crossing of mules with individuals of pure species, and not to the intermixture of hybrid with hybrid.

6thly. From the above considerations, it appears that species have a real existence in nature, and that each was endowed, at the time of its creation, with the attributes and organization by which it is now distinguished.

CHAPTER V.

Laws which regulate the geographical distribution of species—Analogy of climate not attended with identity of species—Botanical geography—Stations—Habitations—Distinct provinces of indigenous plants—Vegetation of islands—Marine vegetation—In what manner plants become diffused—Effects of wind, rivers, marine currents—Agency of animals—Many seeds pass through the stomachs of animals and birds undigested—Agency of man in the dispersion of plants, both voluntary and involuntary—Its analogy to that of the inferior animals.

NEXT to determining the question whether species have a real existence, the consideration of the laws which regulate their geographical distribution is a subject of primary importance to the geologist. It is only by studying these laws with attention, by observing the position which groups of species occupy at present, and inquiring how these may be varied in the course of time by migrations, by changes in physical geography, and other causes, that we can hope to learn whether the duration of species be limited, or in what manner the state of the animate world is affected by the endless vicissitudes of the inanimate.

That different regions of the globe are inhabited by entirely distinct animals and plants is a fact which has been familiar to all naturalists since Buffon first pointed out the want of *specific* identity between the land quadrupeds of America and those of the Old World. The same phenomenon has, in later times, been forced, in a striking manner, upon our attention, by the examination of New Holland, where the indigenous species of animals and plants were found to be, almost without exception, distinct from those known in other parts of the world.

But the extent of this parcelling out of the globe amongst different *nations*, as they have been termed, of plants and animals,—the universality of a phenomenon so extraordinary and unexpected, may be considered as one of the most interesting facts clearly established by the advance of modern science.

Scarcely fourteen hundred species of plants appear to have been known and described by the Greeks, Romans, and Arabians. At present, more than three thousand species are enumerated, as natives of our own island *. In other parts of the world there have been collected, perhaps, upwards of seventy thousand species. It was not to be supposed, therefore, that the ancients should have acquired any correct notions respecting what may be called the geography of plants, although the influence of climate on the character of the vegetation could hardly have escaped their observation.

Antecedently to investigation, there was no reason for presuming that the vegetable productions, growing wild in the eastern hemisphere, should be unlike those of the western, in the same latitude; nor that the plants of the Cape of Good Hope should be unlike those of the South of Europe; situations where the climate is little dissimilar. The contrary supposition would have seemed more probable, and we might have anticipated an almost perfect identity in the animals and plants which inhabit corresponding parallels of latitude. The discovery, therefore, that each separate region of the globe, both of the land and water, is occupied by distinct groups of species, and that most of the exceptions to this general rule may be referred to disseminating causes now in operation, is eminently calculated to excite curiosity, and to stimulate us to seek some hypothesis respecting the first introduction of species which may be reconcileable with such phenomena.

A comparison of the *plants* of different regions of the globe affords results more to be depended upon in the present state of our knowledge, than those relating to the animal kingdom, because the science of botany is more advanced, and probably comprehends a great proportion of the total number of the vegetable productions of the whole earth.

Humboldt, in several eloquent passages of his Personal Narrative, was among the first to promulgate philosophical views on this subject. Every hemisphere, says this traveller,

* Barton's Lectures on the Geography of Plants, p. 2.

produces plants of different species; and it is not by the diversity of climates that we can attempt to explain why equinoctial Africa has no lauriniæ, and the New World no heaths; why the calceolariæ are found only in the southern hemisphere; why the birds of the continent of India glow with colours less splendid than the birds of the hot parts of America; finally, why the tiger is peculiar to Asia, and the ornithorhynchus to New Holland *.

"We can conceive, he adds, that a small number of the families of plants, for instance the musaceæ and the palms, cannot belong to very cold regions, on account of their internal structure and the importance of certain organs; but we cannot explain why no one of the family of melastomas vegetates north of the parallel of thirty degrees; or why no rose-tree belongs to the southern hemisphere. Analogy of climates is often found in the two continents without identity of productions †."

The luminous essay of Decandolle on " Botanical Geography" presents us with the fruits of his own researches and those of Humboldt, Brown, and other eminent botanists, so arranged, that the principal phenomena of the distribution of plants are exhibited in connexion with the causes to which they are chiefly referrible ‡. " It might not, perhaps, be difficult," observes this writer, " to find two points, in the United States and in Europe, or in equinoctial America and Africa, which present all the same circumstances: as for example, the same temperature, the same height above the sea, a similar soil, an equal dose of humidity, yet nearly all, *perhaps all*, the plants in these two similar localities shall be distinct. A certain degree of analogy, indeed, of aspect, and even of structure, might very possibly be discoverable between the plants of the two localities in question, but the *species* would in general be different. Circumstances, therefore, different from those which now determine the *stations*, have had an influence on the *habitations* of plants."

* Pers. Narr., vol. v. p. 180.   † Ibid.

‡ Essai Elémentaire de Géographie Botanique. Extrait du 18e vol. du Dict. des Sci. Nat.

As we shall frequently have occasion to speak of the *stations* and *habitations* of plants in the technical sense in which the terms are used in the above passage, we may remind the geological reader that station indicates the peculiar nature of the locality where each species is accustomed to grow, and has reference to climate, soil, humidity, light, elevation above the sea, and other analogous circumstances ; whereas by habitation is meant a general indication of the country where a plant grows wild. Thus the *station* of a plant may be a salt-marsh, in a temperate climate, a hill-side, the bed of the sea, or a stagnant pool. Its *habitation* may be Europe, North America, or New Holland between the tropics. The study of stations has been styled the topography, that of habitations the geography of botany. The terms thus defined, express each a distinct class of ideas, which have been often confounded together, and which are equally applicable in zoology.

In further illustration of the principle above alluded to, that difference of longitude, independently of any influence of temperature, is accompanied by a great, and sometimes a complete diversity in the species of plants, Decandolle observes, that out of two thousand eight hundred and ninety-one species of phanerogamic plants described by Pursh, in the United States, there are only three hundred and eighty-five which are found in northern or temperate Europe. MM. Humboldt and Bonpland, in all their travels through equinoctial America, found only twenty-four species (these being all cyperacea and graminea) common to America and any part of the Old World. On comparing New Holland with Europe, Mr. Brown ascertained that out of four thousand one hundred species, discovered in Australia, there were only one hundred and sixty-six common to Europe, and of this small number there were some which may have been transported thither by man. Most of the others belong to those classes which are provided with the most ample means of dispersion to vast distances.

But it is still more remarkable, that in the more widely separated parts of the ancient continent, notwithstanding the existence of an uninterrupted land communication, the diversity in the specific character of the respective vegetations is almost as striking. Thus there is found one assemblage of species in China, another in the countries bordering the Black Sea and the Caspian, a third in those surrounding the Mediterranean, a fourth in the great platforms of Siberia and Tartary, and so forth.

The distinctness of the groups of indigenous plants, in the same parallel of latitude, is greatest where continents are disjoined by a wide expanse of ocean. In the northern hemisphere, near the Pole, where the extremities of Europe, Asia and America unite or approach near to one another, a considerable number of the same species of plants are found, common to the three continents. But it has been remarked, that these plants, which are thus so widely diffused in the Arctic regions, are also found in the chain of the Aleutian islands, which stretch almost across from America to Asia, and which may probably have served as the channel of communication for the partial blending of the Floras of the adjoining regions. It has, indeed, been found to be a general rule, that plants found at two points very remote from each other, occur also in places intermediate.

In islands very distant from continents, the total number of plants is comparatively small; but a large proportion of the species are such as occur nowhere else. In so far as the Flora of such islands is not peculiar to them, it contains, in general, species common to the nearest main lands *.

The islands of the great southern ocean exemplify these rules; the easternmost containing more American, and the western more Indian plants †. Madeira and Teneriffe contain many species, and even entire genera, peculiar to them; but they have also plants in common with Portugal, Spain, the Azores, and the north-west coast of Africa ‡.

---

* Prichard, vol. i. p. 36. Brown, Appendix to Flinders.

† Forster, Observations, &c.

‡ Humboldt, Pers. Narr., vol. i. p. 270 of the translation. Prichard, Phys. Hist. of Mankind, vol. i. p. 37.

In the Canaries, out of five hundred and thirty-three species of phanerogamous plants, it is said that three hundred and ten are peculiar to these isles, and the rest identical with those of the African continent; but in the Flora of St. Helena, which is so far distant, even from the western shores of Africa, there have been found, out of sixty-one native species, only *two or three* which are to be found in any other part of the globe.

Decandolle has enumerated twenty great botanical provinces inhabited by indigenous or aboriginal plants; and although many of these contain a variety of species which are common to several others, and sometimes to places very remote, yet the lines of demarcation are, upon the whole, astonishingly well defined*. Nor is it likely that the bearing of the evidence on which these general views are founded will ever be materially affected, since they are already confirmed by the examination of seventy or eighty thousand species of plants.

The entire change of opinion which the contemplation of these phenomena has brought about is worthy of remark. The first travellers were persuaded that they should find, in distant regions, the plants of their own country, and they took a pleasure in giving them the same names. It was some time before this illusion was dissipated ; but so fully sensible did botanists at last become of the extreme smallness of the number of phænogamous plants common to different continents, that the ancient Floras fell into disrepute. All grew diffident of the pretended identifications, and we now find that every naturalist is inclined to examine each supposed exception with scrupulous severity†. If they admit the fact, they begin to speculate on the mode whereby the seeds may have been transported from one country into the other, or inquire on which of two continents the plant was indigenous, assuming that a species, like an individual, cannot have two birth-places.

The marine vegetation is less known, but we learn from

---

* See a farther subdivision 'by which twenty-seven provinces are made, by M. Alph. Decandolle, son of Decandolle. Monogr. des Campanulées. Paris, 1830.

† Decandolle, Essai Elémen. de Géog. Botan. p. 45.

Lamouroux, that it is divisible into different systems, apparently as distinct as those on the land, notwithstanding that the uniformity of temperature is so much greater in the ocean. For on that ground we might have expected the phenomenon of partial distribution to have been far less striking, since climate is, in general, so influential a cause in checking the dispersion of species from one zone to another.

The number of hydrophytes, as they are termed, is very considerable, and their stations are found to be infinitely more varied than could have been anticipated; for while some plants are covered and uncovered daily by the tide, others live in abysses of the ocean, at the extraordinary depth of one thousand feet; and although in such situations there must reign darkness more profound than night, at least to our organs, many of these vegetables are highly coloured. From the analogy of terrestrial plants we should have inferred that the colouring of the algæ was derived from the influence of the solar rays; yet we are compelled to doubt when we reflect how feeble must be the rays which penetrate to these great depths.

The subaqueous vegetation of the Mediterranean is, upon the whole, distinct from that of the Atlantic on the west, and that part of the Arabian gulf which is immediately contiguous on the south. Other botanical provinces are found in the West-Indian seas, including the gulf of Mexico; in the ocean which washes the shores of South America, in the Indian ocean and its gulfs, in the seas of Australia, and in the Atlantic basin, from the 40° of north lat. to the pole. There are very few species common to the coast of Europe and the United States of North America, and none common to the Straits of Magellan and the shores of Van Diemen's Land.

It must not be overlooked, that the distinctness alluded to between the vegetation of these several countries relates strictly to *species* and not to forms. In regard to the numerical preponderance of certain forms, and many peculiarities of internal structure, there is a marked agreement in the vegetable productions of districts placed in corresponding latitudes, and under similar physical circumstances, however remote their position. Thus there are innumerable points of analogy between the vegetation of the Brazils, equinoctial Africa, and India; and there are also points of difference wherein the plants of these regions are distinguishable from all extra-tropical groups. But there are very few species common to the three continents. The same may be said, if we compare the plants of the Straits of Magellan with those of Van Diemen's Land, or the vegetation of the United States with that of the middle of Europe: the species are distinct, but the forms are in a great degree analogous.

Let us now consider what means of diffusion, independently of the agency of man, are possessed by plants, whereby, in the course of ages, they may be enabled to stray from one of the botanical provinces above mentioned to another, and to establish new colonies at a great distance from their birth-place.

The principal of the inanimate agents, provided by nature for scattering the seeds of plants over the globe, are the movements of the atmosphere and of the ocean, and the constant flow of water from the mountains to the sea. To begin with the winds: a great number of seeds are furnished with downy and feathery appendages, enabling them, when ripe, to float in the air, and to be wafted easily to great distances by the most gentle breeze. Other plants are fitted for dispersion by means of an attached wing, as in the case of the fir-tree, so that they are caught up by the wind as they fall from the cone, and are carried to a distance. Amongst the comparatively small number of plants known to Linnæus, no less than one hundred and thirty-eight genera are enumerated as having winged seeds.

As winds often prevail for days, weeks, or even months together, in the same direction, these means of transportation may sometimes be without limits; and even the heavier grains may be borne through considerable spaces, in a very short time, during ordinary tempests; for strong gales, which can sweep along grains of sand, often move at the rate of about forty miles an hour, and if the storm be very violent, at the rate of fifty-six miles *. The hurricanes of tropical regions, which root up trees and throw down buildings, sweep along at the rate of ninety miles an hour, so that, for however short a time they prevail, they may carry even the heavier fruits and seeds over friths and seas of considerable width, and, doubt-less, are often the means of introducing into islands the vegetation of adjoining continents. Whirlwinds are also in-strumental in bearing along heavy vegetable substances to considerable distances. Slight ones may frequently be observed in our fields, in summer, carrying up haycocks into the air, and then letting fall small tufts of hay far and wide over the country; but they are sometimes so powerful as to dry up lakes and ponds, and to break off the boughs of trees, and carry them up in a whirling column of air.

Franklin tells us, in one of his letters, that he saw, in Maryland, a whirlwind which began by taking up the dust which lay in the road, in the form of a sugar-loaf with the pointed end downwards, and soon after grew to the height of forty or fifty feet, being twenty or thirty in diameter. It advanced in a direction contrary to the wind, and although the rotatory motion of the column was surprisingly rapid, its onward progress was sufficiently slow to allow a man to keep pace with it on foot. Franklin followed it on horseback, accompanied by his son, for three-quarters of a mile, and saw it enter a wood, where it twisted and turned round large trees with surprising force. These were carried up in a spiral line, and were seen flying in the air, together with boughs and innumerable leaves, which, from their height, appeared reduced to the apparent size of flies. As this cause operates at different intervals of time throughout a great portion of the earth's surface, it may be the means of bearing not only plants but insects, land-testacea and their eggs, with many other species of animals, to points which they could never otherwise

* Annuaire du Bureau des Longitudes.

have reached, and from which they may then begin to propagate themselves again as from a new centre.

The seeds of some aquatic fresh-water plants are of the form of shells, or small canoes, and on this account they swim on the surface, and are carried along by the wind and stream. Others are furnished with fibres, which serve the purpose of masts and sails, so that they are impelled along by the winds, even where there is no current. They cannot take root until the water stagnates, or till they reach some sheltered corner, where they may live without being exposed to too much agitation from winds and currents *. The above-mentioned contrivances may enable aquatic plants to diffuse themselves gradually to considerable distances wherever there is a great chain of lakes, or a river which traverses a large continent.

It has been found that a great numerical proportion of the exceptions to the limitation of species to certain quarters of the globe, occur in the various tribes of cryptogamic plants. Linnæus observed, that as the germs of plants of this class, such as mosses, fungi, and lichens, consist of an impalpable powder, the particles of which are scarcely visible to the naked eye, there is no difficulty to account for their being dispersed throughout the atmosphere, and carried to every point of the globe, where there is a station fitted for them. Lichens in particular ascend to great elevations, sometimes growing two thousand feet above the line of perpetual snow, at the utmost limits of vegetation, and where the mean temperature is nearly at the freezing point. This elevated position must contribute greatly to facilitate the dispersion of those buoyant particles of which their fructification consists †.

Some have inferred, from the springing up of mushrooms whenever particular soils and decomposed organic matter are mixed together, that the production of fungi is accidental, and not analogous to that of perfect plants ‡. But Fries, whose authority on these questions is entitled to the highest respect,

* Rev. Dr. Rennie, Essays on the Nat. Hist. of Peat Moss, p. 248.
† Linn., Tour in Lapland, vol. ii. p. 282.
‡ Lindley, Introd. to Nat. Syst. of Botany, who cites Fries.

has shown the fallacy of this argument in favour of the old
doctrine of equivocal generation. " The sporules of fungi,"
says this naturalist, " are so infinite, that in a single individual
of Reticularia maxima, I have counted above ten millions, and
so subtile as to be scarcely visible, often resembling thin smoke;
so light that they may be raised perhaps by evaporation into
the atmosphere, and dispersed in so many ways by the attrac-
tion of the sun, by insects, wind, elasticity, adhesion, &c., that
it is difficult to conceive a place from which they may be excluded."

In turning our attention, in the next place, to the instrumentality of the aqueous agents of dispersion, we cannot do
better than cite the words of one of our ablest botanical writers.
" The mountain-stream or torrent," observes Keith, " washes
down to the valley the seeds which may accidentally fall into
it, or which it may happen to sweep from its banks when it
suddenly overflows them. The broad and majestic river,
winding along the extensive plain, and traversing the continents
of the world, conveys to the distance of many hundreds of miles
the seeds that may have vegetated at its source. Thus the
southern shores of the Baltic are visited by seeds which grew
in the interior of Germany; and the western shores of the
Atlantic by seeds that have been generated in the interior of
America *." Fruits, moreover, indigenous to America and
the West Indies, such as that of the Mimosa scandens, the
cashew-nut, and others, have been known to be drifted across
the Atlantic by the Gulf-stream, on the western coasts of
Europe, in such a state that they might have vegetated had
the climate and soil been favourable. Among these the Gui-
landina Bonduc, a leguminous plant, is particularly mentioned,
as having been raised from a seed found on the west coast of
Ireland †. Sir Hans Sloane informs us that the lenticula
marina, or sargasso, a bean which is frequently cast ashore on
the Orkney isles, and coast of Ireland, grows on the rocks
about Jamaica, where the surface of the sea is sometimes

---

* System of Physiological Botany, vol. ii. p. 405.
† Brown, Append. to Tuckey, No. V. p. 481.

strewed with it, and from whence it is known to be carried by the winds and currents towards the coast of Florida *.

The absence of liquid matter in the composition of seeds renders them comparatively insensible to heat and cold, so that they may be carried, without detriment, through climates where the plants themselves would instantly perish.   Such is their power of resisting the effects of heat, that Spallanzani mentions some seeds that germinated after having been boiled in water †.   When, therefore, a strong gale, after blowing violently off the land for a time, dies away, and the seeds alight upon the surface of the waters, or wherever the ocean, by eating away the sea-cliffs, throws down into its waves plants which would never otherwise approach the shores, the tides and currents become active instruments in assisting the dissemination of almost all classes of the vegetable kingdom.

In a collection of six hundred plants from the neighbourhood of the river Zaire, in Africa, Mr. Brown found that thirteen species were also met with on the opposite shores of Guiana and Brazil.   He remarked, that most of these plants were only found on the lower parts of the river Zaire, and were chiefly such as produced seeds capable of retaining their vitality a long time in the currents of the ocean.

Islands, moreover, and even the smallest rocks, play an important part in aiding such migrations, for when seeds alight upon them from the atmosphere, or are thrown up by the surf, they often vegetate and supply the winds and waves with a repetition of new and uninjured crops of fruits and seeds, which may afterwards pursue their course through the atmosphere, or along the surface of the sea, in the same direction.   The number of plants found at any given time on an islet affords no test whatever of the extent to which it may have co-operated towards this end, since a variety of species may first thrive there and then perish, and be followed by other chance-comers like themselves.

Currents and winds, in the arctic regions, drift along ice-

* Phil. Trans. 1696.   † System of Philosophical Botany, vol. ii. p. 403,

bergs covered with an alluvial soil on which herbs and pine saplings are seen growing, which often continue to vegetate on some distant shore where the ice-island is stranded.

With respect to marine vegetation, the seeds being in their native element, may remain immersed in water without injury for indefinite periods, so that there is no difficulty in conceiving the diffusion of species wherever uncongenial climates, contrary currents, and other causes, do not interfere. All are familiar with the sight of the floating sea-weed

" Flung from the rock on ocean's foam to sail,
  Where'er the surge may sweep, the tempest's breath prevail."

Remarkable accumulations of drift weed occur on each side of the equator in the Atlantic, Pacific, and Indian Oceans. Columbus and other navigators who first encountered these banks of algæ in the Northern Atlantic, compared them to vast inundated meadows, and state that they retarded the progress of their vessels. The most extensive bank is a little west of the meridian of Fayal, one of the Azores, between latitude 25° and 36°; violent north winds sometimes prevail in this space, and drive the sea-weed to low latitudes, as far as the 24th or even the 20th degree*.

The hollow pod-like receptacles in which the seeds of many algæ are lodged, and the filaments attached to the seed-vessels of others, seem intended to give buoyancy, and we may observe that these hydrophytes are in general *proliferous*, so that the smallest fragment of a branch can be developed into a perfect plant. The seeds, moreover, of the greater number of species are enveloped with a mucous matter like that which surrounds the eggs of some fish, and which not only protects them from injury, but serves to attach them to floating bodies or to rocks.

But we have as yet considered part only of the fertile resources of nature for conveying seeds to a distance from their place of growth. The various tribes of animals are busily engaged in furthering an object whence they derive such important advantages. Sometimes an express provision is found

* Greville, Introduction to Algæ Britannicæ, p. 12.

in the structure of seeds to enable them to adhere firmly by prickles, hooks, and hairs, to the coats of animals, or feathers of the winged tribe, to which they remain attached for weeks, or even months, and are borne along into every region whither birds or quadrupeds may migrate. Linnæus enumerates fifty genera of plants, and the number now known to botanists is much greater, which are armed with hooks by which, when ripe, they adhere to the coats of animals. Most of these vegetables, he remarks, require a soil enriched with dung. Few have failed to mark the locks of wool hanging on the thorn-bushes, wherever the sheep pass, and it is probable that the wolf or lion never give chace to herbivorous animals without being unconsciously subservient to this part of the vegetable economy.

A deer has strayed from the herd, when browsing on some rich pasture, when he is suddenly alarmed by the approach of his foe. He instantly plunges through many a thicket, and swims through many a river and lake. The seeds of the herbs and shrubs adhere to his smoking flanks, and are washed off again by the streams. The thorny spray is torn off and fixes itself in his hairy coat, until brushed off again in other thickets and copses. Even on the spot where the victim is devoured, many of the seeds which he had swallowed immediately before the pursuit may be left on the ground uninjured.

The passage, indeed, of undigested seeds through the stomachs of animals is one of the most efficient causes of the dissemination of plants, and is of all others, perhaps, the most likely to be overlooked. Few are ignorant that a portion of the oats eaten by a horse preserve their germinating faculty in the dung. The fact of their being still nutritious is not lost on the sagacious rook. To many, says Linnæus, it seems extraordinary, and something of a prodigy, that when a field is well tilled and sown with the best wheat, it frequently produces darnel or the wild oat, especially if it be manured with new dung : they do not consider that the fertility of the smaller seeds is not destroyed in the ventricles of animals *.

* Linnæus, Amœn. Acad., vol. ii. p. 409.

Some of the order of the Passeres, says Ekmarck *, devour the seeds of plants in great quantities, which they eject again in very distant places, without destroying its faculty of vegetation ; thus a flight of larks will fill the cleanest field with a great quantity of various kinds of plants, as the melilot trefoil (*Medicago lupulina*), and others whose seeds are so heavy that the wind is not able to scatter them to any distance. In like manner, the blackbird and missel-thrush, when they devour berries in too great quantities, are known to consign them to the earth undigested in their excrement †.

Pulpy fruits serve quadrupeds and birds as food, while their seeds, often hard and indigestible, pass uninjured through the intestines, and are deposited far from their original place of growth in a condition peculiarly fit for vegetation ‡. So well are our farmers, in some parts of England, aware of this fact, that when they desire to raise a quick-set hedge in the shortest possible time, they feed turkeys with the haws of the common white-thorn (*Cratægus oxyacantha*), and then sow the stones which are ejected in their excrement, whereby they gain an entire year in the growth of the plant §. Birds when they pluck cherries, sloes, and haws, fly away with them to some convenient place, and when they have devoured the fruit drop the stone into the ground. Captain Cook, in his account of the volcanic island of Tanna, one of the New Hebrides, which he visited in his second voyage, makes the following interesting observation. " Mr. Forster, in his botanical excursion this day, shot a pigeon, in the craw of which was a wild nutmeg. He took some pains to find the tree on this island, but his endeavours were without success ||." It is easy, therefore, to perceive, that birds in their migrations to great distances, and even across seas, may transport seeds to new isles and continents.

---

* Amœn. Acad., vol. iv., Essay 75, § 8.

† Wilcke, Amœn. Acad., vol. vi. § 22.

‡ Smith's Introd. to Phys. and Syst. Botany, p. 304, 1807.

§ This information was communicated to me by Professor Henslow, of Cambridge. || Book iii., ch. 4.

The sudden deaths to which great numbers of frugivorous birds are annually exposed, must not be omitted as auxiliary to the transportation of seeds to new habitations. When the sea retires from the shore, and leaves fruits and seeds on the beach, or in the mud of estuaries, it might, by the returning tide, wash them away again, or destroy them by long immersion; but when they are gathered by land birds which frequent the sea-side, or by waders and water-fowl, they are often borne inland, and if the bird to whose crop they have been consigned is killed, they may be left to grow up far from the sea. Let such an accident happen but once in a century, or a thousand years, it will be sufficient to spread many of the plants from one continent to another; for, in estimating the activity of these causes, we must not consider whether they act slowly in relation to the period of our observation, but in reference to the duration of species in general.

Let us trace the operation of this cause in connexion with others. A tempestuous wind bears the seeds of a plant many miles through the air, and then delivers them to the ocean; the oceanic current drifts them to a distant continent; by the fall of the tide they become the food of numerous birds, and one of these is seized by a hawk or eagle, which, soaring across hill and dale to a place of retreat, leaves, after devouring its prey, the unpalatable seeds to spring up and flourish in a new soil.

The machinery before adverted to is so capable of disseminating seeds over almost unbounded spaces, that were we more intimately acquainted with the economy of nature, we might probably explain all the instances which occur of the aberration of plants to great distances from their native countries. The real difficulty which must present itself to every one who contemplates the present geographical distribution of species, is the small number of exceptions to the rule of the non-intermixture of different groups of plants. Why have they not, supposing them to have been ever so distinct originally, become more blended and confounded together in the lapse of ages?

But in addition to all the agents already enumerated as instrumental in diffusing plants over the globe, we have still to consider man—one of the most important of all. He transports with him, into every region, the vegetables which he cultivates for his wants, and is the involuntary means of spreading a still greater number which are useless to him, or even noxious. " When the introduction of cultivated plants is of recent date, there is no difficulty in tracing their origin ; but when it is of high antiquity, we are often ignorant of the true country of the plants on which we feed. No one contests the American origin of the maize or the potato, nor the origin, in the old world, of the coffee-tree and of wheat. But there are certain objects of culture, of very ancient date, between the tropics, such, for example, as the banana, of which the origin cannot be verified. Armies, in modern times, have been known to carry, in all directions, grain and cultivated vegetables from one extremity of Europe to the other, and thus have shown us how, in more ancient times, the conquests of Alexander, the distant expeditions of the Romans, and afterwards the Crusades, may have transported many plants from one part of the world to the other *."

But besides the plants used in agriculture, the number which have been naturalized by accident, or which man has spread unintentionally, is considerable. One of our old authors, Josselyn, gives a catalogue of such plants as had, in his time, sprung up in the colony since the English planted and kept cattle in New England. They were two and twenty in number. The common nettle was the first which the settlers noticed, and the plantain was called by the Indians, " English man's foot," as if it sprung from their footsteps †.

" We have introduced everywhere," observes Decandolle, " some weeds which grow among our various kinds of wheat, and which have been received, perhaps, originally from Asia with them. Thus, together with the Barbary wheat, the inhabi-

* Decandolle, Essai Elémen. &c. p. 50.
† Quarterly Review, vol. xxx., p. 8.

tants of the south of Europe have sown, for many ages,the plants of Algiers and Tunis. With the wools and cottons of the East, or of Barbary, there are often brought into France, the grains of exotic plants, some of which naturalize themselves. Of this I will cite a striking example. There is at the gate of Montpelier, a meadow set apart for drying foreign wool *after it has been washed.* There hardly passes a year without some foreign plants being found naturalized in this drying ground. I have gathered there Centaurea parviflora, Psoralea palæstina, and Hypericum crispum." This fact is not only illustrative of the aid which man lends inadvertently to the propagation of plants, but it also demonstrates the multiplicity of seeds which are borne about in the woolly and hairy coats of wild animals.

The same botanist mentions instances of plants naturalized in sea-ports by the ballast of ships, and several examples of others which have spread through Europe from botanical gardens, so as to have become more common than many indigenous species.

It is scarcely a century, says Linnæus *, since the Canadian erigiron, or flea-bane, was brought from America to the botanical garden at Paris, and already the seeds have been carried by the winds, so that it is diffused over France, the British islands, Italy, Sicily, Holland, and Germany. Several others are mentioned by the Swedish naturalist, as having been dispersed by similar means. The common thorn-apple, *Datura stramonium*, observes Willdenow, now grows as a noxious weed throughout all Europe, with the exception of Sweden, Lapland, and Russia. It came from the East Indies and Abyssinia to us, and was so universally spread by certain quacks who used its seed as an emetic †.

In hot and ill-cultivated countries, such naturalizations take place more easily. Thus the Chenopodium ambrosioides, sown by Mr. Burchell on a point of St. Helena, multiplied so

---

* Essay on the Habitable Earth, Amœn. Acad. vol. ii. p. 409.
† Principles of Botany, p. 389.

in four years as to become one of the commonest weeds in the island *.

The most remarkable proof, says Decandolle, of the extent to which man is unconsciously the instrument of dispersing and naturalizing species, is found in the fact, that in New Holland, America, and the Cape of Good Hope, the aboriginal European species exceed in number all the others which have come from any distant regions, so that, in this instance, the influence of man has surpassed that of all the other causes which tend to disseminate plants to remote districts.

Although we are but slightly acquainted, as yet, with the extent of our instrumentality in naturalizing species, yet the facts ascertained afford no small reason to suspect that the number which we introduce unintentionally, exceeds all those transported by design. Nor is it unnatural to suppose that the functions, which the inferior beings extirpated by man once discharged in the economy of nature, should devolve upon the human race. If we drive many birds of passage from different countries, we are probably required to fulfil their office of carrying seeds, eggs of fish, insects, molluscs, and other creatures, to distant regions; if we destroy quadrupeds, we must replace them, not merely as consumers of the animal and vegetable substances which they devoured, but as disseminators of plants, and of the inferior classes of the animal kingdom. We do not mean to insinuate that the same changes which man brings about, would have taken place by means of the agency of other species, but merely that he supersedes a certain number of agents, and so far as he disperses plants unintentionally, or against his will, his intervention is strictly analogous to that of the species so extirpated.

We may observe, moreover, that if, at former periods, the animals inhabiting any given district have been partially altered by the extinction of some species, and the introduction of others, whether by new creations or by immigration, a change must have taken place in regard to the particular plants con-

* Principles of Botany, p. 389.

veyed about with them to foreign countries. As for example, when one set of migratory birds is substituted for another, the countries from and to which seeds are transported are immediately changed. Vicissitudes, therefore, analogous to those which man has occasioned, may have previously attended the springing up of new relations between species in the vegetable and animal worlds.

It may also be remarked, that if man is the most active agent in enlarging, so also is he in circumscribing the geographical boundaries of particular plants. He promotes the migration of some, he retards that of other species, so that while in many respects he appears to be exerting his power to blend and confound the various provinces of indigenous species, he is, in other ways, instrumental in obstructing the fusion into one group of the inhabitants of contiguous provinces.

Thus, for example, when two botanical regions exist in the same great continent, such as *the European region,* comprehending the central parts of Europe and those surrounding the Mediterranean, and *the Oriental region,* as it has been termed, embracing the countries adjoining the Black Sea and the Caspian, the interposition between these of thousands of squares miles of cultivated lands, opposes a new and powerful barrier against the mutual interchange of indigenous plants. Botanists are well aware that garden plants naturalize and diffuse themselves with great facility in comparatively unreclaimed countries, but spread themselves slowly and with difficulty in districts highly cultivated. There are many obvious causes for this difference; by drainage and culture the natural variety of stations is diminished, and those stray individuals by which the passage of a species from one fit station to another is effected, are no sooner detected by the agriculturist, than they are uprooted as weeds. The larger shrubs and trees, in particular, can scarcely ever escape observation, when they have attained a certain size, and will rarely fail to be cut down if unprofitable.

The same observations are applicable to the interchange of

the insects, birds, and quadrupeds of two regions situated like those above alluded to. No beasts of prey are permitted to make their way across the intervening arable tracts. Many birds, and hundreds of insects, which would have found some palatable food amongst the various herbs and trees of the primeval wilderness, are unable to subsist on the olive, the vine, the wheat, and a few trees and grasses favoured by man. In addition, therefore, to his direct intervention, man, in this case, operates indirectly to impede the dissemination of plants, by intercepting the migrations of animals, many of which would otherwise have been active in transporting seeds from one province to another.

Whether in the vegetable kingdom the influence of man will tend, after a considerable lapse of ages, to render the geographical range of *species in general* more extended, as Decandolle seems to anticipate, or whether the compensating agency above alluded to will not counterbalance the exceptions caused by our naturalizations, admits at least of some doubt. In the attempt to form an estimate on this subject, we must be careful not to underrate, or almost overlook, as some appear to have done, the influence of man in checking the diffusion of plants, and restricting their distribution to narrower limits.

CHAPTER VI.

Geographical Distribution of Animals—Buffon on the specific distinctness of the quadrupeds of the old and new world—Different regions of indigenous mammalia—Quadrupeds in islands—Range of the Cetacea—Dissemination of quadrupeds—their powers of swimming—Migratory instincts—Drifting of quadrupeds on ice-floes—On floating islands of drift timber—Migrations of Cetacea—Habitations of Birds—Their migrations and facilities of diffusion—Distribution of Reptiles and their powers of dissemination.

ALTHOUGH in speculating on "philosophical possibilities," said Buffon, the same temperature might have been expected, all other circumstances being equal, to produce the same beings in different parts of the globe, both in the animal and vegetable kingdoms, yet it is an undoubted fact, that when America was discovered, its indigenous quadrupeds were all dissimilar from those previously known in the old world. The elephant, the rhinoceros, the hippopotamus, the cameleopard, the camel, the dromedary, the buffalo, the horse, the ass, the lion, the tiger, the apes, the baboons, and a number of other mammalia, where nowhere to be met with on the new continent; while in the old, the American species, of the same great class, were nowhere to be seen—the tapir, the lama, the pecari, the jaguar, the couguar, the agouti, the paca, the coati, and the sloth.

These phenomena, although few in number relatively to the whole animate creation, were so striking and so positive in their nature, that the French naturalist caught sight at once of a general law in the geographical distribution of organic beings, namely, the limitation of groups of distinct species to regions separated from the rest of the globe by certain natural barriers. It was, therefore, in a truly philosophical spirit that, relying on the clearness of the evidence obtained respecting the larger quadrupeds, he ventured to call in question the identifications announced by some contemporary naturalists, of species of animals said to be common to the southern extremities of America and Africa *.

The migration of quadrupeds from one part of the globe to the other, observes one of our ablest writers, is prevented by uncongenial climates and the branches of the ocean which intersect continents. " Hence by a reference to the geographical site of countries, we may divide the earth into a certain number of regions fitted to become the abodes of particular groups of animals, and we shall find, on inquiry, that each of these provinces, thus conjecturally marked out, is actually inhabited by a distinct nation of quadrupeds †."

Where the continents of the old and new world approximate to each other towards the north, the narrow straits which separate them are frozen over in winter, and the distance is further lessened by intervening islands. Thus a passage from one continent to another becomes practicable to such quadrupeds as are fitted to endure the intense cold of the arctic circle. Accordingly, the whole arctic region has become one of the provinces of the animal kingdom, and contains many species common to both the great continents. But the temperate regions of America, which are separated by a wide extent of ocean from those of Europe and Asia, contain each a distinct nation of indigenous quadrupeds. There are three groups of *tropical* mammalia belonging severally to America, Africa, and continental India, each inhabiting lands separated from each other by the ocean.

In Peru and Chili, says Humboldt, the region of the grasses, which is at an elevation of from twelve thousand three hundred to fifteen thousand four hundred feet, is inhabited by crowds of lama, guanaco, and alpaca. These quadrupeds, which here represent the genus camel of the ancient continent,

---

* Buffon, vol. v.—On the Virginian Opossum.

† Prichard's Phys. Hist. of Mankind, vol. i. p. 54. In some of the preliminary chapters will be found a sketch of the leading facts illustrative of the geographical distribution of animals, drawn up with the author's usual clearness and ability.

have not extended themselves either to Brazil or Mexico, because, during their journey, they must necessarily have descended into regions that were too hot for them *.

New Holland is well known to contain a most singular and characteristic assemblage of mammiferous animals, consisting of more than forty species of the marsupial family, of which no congeners even occur elsewhere, with the exception of a few American opossums. This exclusive occupation of the Australian continent by the kangaroos and other tribes of pouched animals, although it has justly excited great attention, is a fact, nevertheless, in strict accordance with the general laws of the distribution of species ; since, in other parts of the globe, we find peculiarities of form, structure, and habit, in birds, reptiles, insects, or plants, confined entirely to one hemisphere, or one continent, and sometimes to much narrower limits.

The southern region of Africa, where that continent extends into the temperate zone, constitutes another separate zoological province, surrounded as it is on three sides by the ocean, and cut off from the countries of milder climate, in the northern hemisphere, by the intervening torrid zone. In many instances, this region contains the same genera which are found in temperate climates to the northward of the line ; but then the southern are different from the northern species. Thus in the south we find the quagga and the zebra : in the north, the horse, the ass, and the jiggetai of Asia.

The south of Africa is spread out into fine level plains from the tropic to the Cape ; in this region, says Pennant, besides the horse genus, of which five species have been found, there are also peculiar species of rhinoceros, the hog, and the hyrax, among pachydermatous races ; and amongst the ruminating, the giraffe, the Cape buffalo, and a variety of remarkable antelopes, as the springbok, the oryx, the gnou, the leucophoë, the pygarga, and several others †.

---

* Description of the Equatorial Regions.

† Pennant's Hist. of Quadrupeds, cited by Prichard, Phys. Hist. of Mankind, vol. i. p. 66.

The Indian archipelago presents peculiar phenomena in regard to its indigenous mammalia, which, in their generic character, recede in some respects from that of the animals of the Indian continent, and approximate to the African. The Sunda isles contain a hippopotamus, which is wanting in the rivers of Asia; Sumatra, a peculiar species of tapir, and a rhinoceros resembling the African more than the Indian species, but specifically distinguishable from both *.

Beyond the Indian archipelago is an extensive region, including New Guinea, New Britain, and New Ireland, together with the archipelago of Solomon's islands, the New Hebrides, and Louisiade, and the more remote groups of isles in the great southern ocean, which may be considered as forming one zoological province. Although these remarkable countries are extremely fertile in their vegetable productions, they are almost wholly destitute of native warm-blooded quadrupeds, except a few species of bats, and some domesticated animals in the possession of the natives †.

Quadrupeds found on islands situated near the continents, generally form a part of the stock of animals belonging to the adjacent main land; "but small islands remote from continents are in general altogether destitute of land quadrupeds, except such as appear to have been conveyed to them by men. Kerguelen's Land, Juan Fernandez, the Gallapagos, and the Isles de Lobos, are examples of this fact. Among all the groups of fertile islands in the Pacific ocean, no quadrupeds have been found, except dogs, hogs, rats, and a few bats. The bats have been found in New Zealand and the more westerly groups; they may probably have made their way along the chain of islands which extend from the shores of New Guinea far into the southern Pacific. The hogs and the dogs appear to have been conveyed by the natives from New Guinea. The Indian isles, near New Guinea, abound in oxen, buffaloes, goats, deer, hogs, dogs, cats, and rats; but none of them are said to have

---

* Prichard, Phys. Hist. of Mankind, p. 66; Cuvier, Ann. du Museum, tom. vii.
         † Prichard, ibid., p. 56.

reached New Guinea, except the hog and the dog. The New Guinea hog is of the Chinese variety, and was probably brought from some of the neighbouring isles, being the animal most in request among savages. It has run wild in New Guinea. Thence it has been conveyed to the New Hebrides, the Tonga and Society isles, and to the Marquesas ; but it is still wanting in the more easterly islands, and, to the southward, in New Caledonia.

"Dogs may be traced from New Guinea to the New Hebrides and Fiji isles ; but they are wanting in the Tonga isles, though found among the Society and Sandwich islanders, by some of whom they are used for food : to the southward they have been conveyed to New Caledonia and New Zealand. In Easter Island, the most remotely situated in this ocean, there are no domestic animals except fowls and rats, which are eaten by the natives : these animals are found in most of the islands ; the fowls are probably from New Guinea. Rats are to be found even on some desert islands, whither they may have been conveyed by canoes which have occasionally approached the shores. It is known, also, that rats occasionally swim in large numbers to considerable distances *."

It is natural to suppose that the geographical range of the different species of cetacea should be less correctly ascertained than that of the terrestrial mammifers. It is, however, well known, that the whales which are obtained by our fishers in the South Seas, are distinct from those of the North ; and the same dissimilarity has been found in all the other marine animals of the same class, so far as they have yet been studied by naturalists.

Let us now inquire what facilities the various land quadrupeds enjoy of spreading themselves over the surface of the earth. In the first place, as their numbers multiply, all of them, whether they feed on plants, or prey on other animals, are disposed to scatter themselves gradually over as wide an area as is accessible to them. But before they have extended their

---

* Prichard, Phys. Hist. of Mankind, vol. i., p. 75.

migrations over a large space, they are usually arrested either by the sea, or a zone of uncongenial climate, or some lofty and unbroken chain of mountains, or a tract already occupied by a hostile and more powerful species.

Rivers and narrow friths can seldom interfere with their progress, for the greater part of them swim well, and few are without this power when urged by danger and pressing want. Thus, amongst beasts of prey, the tiger is seen swimming about among the islands and creeks in the delta of the Ganges, and the jaguar traverses with ease the largest streams in South America *. The bear, also, and the bison, stem the current of the Mississippi. The popular error, that the common swine cannot escape by swimming when thrown into the water, has been contradicted by several curious and well-authenticated instances during the recent floods in Scotland. One pig, only six months old, after having been carried down from Garmouth to the bar at the mouth of the Spey, a distance of a quarter of a mile, swam four miles eastward to Port Gordon and landed safe. Three others, of the same age and litter, swam at the same time five miles to the west, and landed at Blackhill †.

In an adult and wild state, these animals would doubtless have been more strong and active, and might, when hard pressed, have performed a much longer voyage. Hence islands remote from the continent may obtain inhabitants by casualties which, like the late storms in Morayshire, may only occur once in many centuries, or thousands of years, under all the same circumstances. It is obvious that powerful tides, winds, and currents, may sometimes carry along quadrupeds capable, in like manner, of preserving themselves for hours in the sea to very considerable distances, and in this way, perhaps, the tapir (*Tapir Indicus*) may have become common to Sumatra and the Malayan peninsula.

To the elephant in particular, the power of crossing rivers is essential in a wild state, for the quantity of food which a herd

* Buffon, vol. v. p. 204.

† Sir T. D. Lauder, Bart. on the Floods in Morayshire, August, 1829, p. 302, second edition.

of these animals consumes renders it necessary that they should be constantly moving from place to place. The elephant crosses the stream in two ways. If the bed of the river be hard, and the water not of too great a depth, he fords it. But when he crosses great rivers, such as the Ganges and the Niger, the elephant swims deep, so deep that the end of his trunk only is out of the water* ; for it is a matter of indifference to him whether his body be completely immersed, provided he can bring the tip of his trunk to the surface, so as to breathe the external air.

Animals of the deer kind frequently take to the water, especially in the rutting season, when the stags are seen swimming for several leagues at a time, from island to island, in search of the does, especially in the Canadian lakes ; and in some countries where there are islands near the sea-shore, they fearlessly enter the sea and swim to them. In hunting excursions, in North America, the elk of that country is frequently pursued for great distances through the water.

The large herbivorous animals, which are gregarious, can never remain long in a confined region, as they consume so much vegetable food. The immense herds of bisons which often, in the great valley of the Mississippi, blacken the surface near the banks of that river and its tributaries, are continually shifting their quarters, followed by wolves which prowl about in their rear. " It is no exaggeration," says Mr. James, " to assert, that in one place, on the banks of the Platte, at least ten thousand bisons burst on our sight in an instant. In the morning we again sought the living picture, but upon all the plain, which last evening was so teeming with noble animals, not one remained †."

Besides the disposition common to the individuals of every species slowly to extend their range in search of food, in proportion as their numbers augment, a migratory instinct often develops itself in an extraordinary manner, when, after an

* Lib. Entert. Know., Quadrupeds, vol. ii. p. 63.
† Expedition from Pittsburgh to the Rocky Mountains, vol. ii. p. 153.

unusually prolific season, or upon a sudden scarcity of provisions, great multitudes are threatened by famine. We shall
enumerate several illustrations of these migrations, because
they may put us upon our guard against attributing a high
antiquity to a particular species merely because it is diffused
over a great space ; they show clearly how soon, in a state of
nature, a newly-created species might spread itself, in every
direction, from a single point.

In very severe winters, great numbers of the black bears of
America migrate from Canada into the United States ; but
in milder seasons, when they have been well fed, they remain
and hybernate in the north *. The rein-deer, which in Scandinavia can scarcely exist to the south of the sixty-fifth parallel,
descends, in consequence of the greater coldness of the climate,
to the fiftieth degree, in Chinese Tartary, and often roves
into a country of more southern latitude than any part of
England.

In Lapland, and other high latitudes, the common squirrels,
whenever they are compelled, by want of provisions, to quit
their usual abodes, migrate in amazing numbers, and travel
directly forwards, allowing neither rocks, forests, nor the
broadest waters, to turn them from their course. Great numbers are often drowned in attempting to pass friths and rivers.
In like manner the small Norway rat sometimes pursues its
migrations in a straight line across rivers and lakes ; and Pennant informs us, that when, in Kamtschatka, the rats become
too numerous, they gather together in the spring, and proceed
in great bodies westward, swimming over rivers, lakes, and
arms of the sea. Many are drowned or destroyed by water-
fowl or fish. As soon as they have crossed the river Penchim,
at the head of the gulf of the same name, they turn southward,
and reach the rivers Judoma and Ochot by the middle of July,
a district surprisingly distant from their point of departure.

The lemings, also, of Scandinavia, often pour down in
myriads from the northern mountains and devastate the coun-

* Richardson's Fauna Boreali-Americana, p. 16.

try. They generally move in lines which are about three feet
from each other, and exactly parallel, and they direct their
march from the north-west to the south-east, going directly
forward through rivers and lakes, and when they meet with
stacks of hay or corn, gnawing their way through them instead
of passing round *.

Vast troops of the wild ass, or *onager* of the ancients, which
inhabit the mountainous deserts of Great Tartary, feed, during
the summer, in the tracts east and north of Lake Aral. In the
autumn they collect in herds of hundreds, and even thousands,
and direct their course towards the north of India, and often to
Persia, to enjoy a warm retreat during winter †. Bands of two
or three hundred quaggas, a species of wild ass, are sometimes seen to migrate from the tropical plains of southern Africa
to the vicinity of the Malaleveen river. During their migrations they are followed by lions, who slaughter them night by
night ‡.

The migratory swarms of the springbok, or Cape antelope,
afford another illustration of the rapidity with which a species,
under certain circumstances, may be diffused over a continent.
When the stagnant pools of the immense deserts south of the
Orange river dry up, which often happens after intervals of
three or four years, myriads of these animals desert the parched
soil, and pour down like a deluge on the cultivated regions
nearer the Cape. The havoc committed by them resembles
that of the African locusts ; and so crowded are the herds, that
" the lion has been seen to walk in the midst of the compressed
phalanx with only as much room between him and his victims
as the fears of those immediately around could procure by
pressing outwards §."

Dr. Horsfield mentions a singular fact in regard to the geo-

---

* Phil. Trans., vol. ii. p. 872.   † Wood's Zoography, vol. i. p. 11.
‡ On the authority of Mr. Campbell. Library of Entert. Know., Menageries,
vol. i. p. 152.
§ Cuvier's Animal Kingdom, by Griffiths, vol. ii. p. 109. Library of Entert.
Know., Menageries, vol. i. p. 366.

graphical distribution of the Mydaus meliceps, a kind of polecat inhabiting Java. This animal is " confined exclusively to
those mountains which have an elevation of more than seven
thousand feet above the level of the ocean : on these it occurs
with the same regularity as many plants. The long-extended
surface of Java, abounding with conical points which exceed
this elevation, affords many places favourable for its resort.
On ascending these mountains, the traveller scarcely fails to
meet with this animal, which, from its peculiarities, is universally known to the inhabitants of these elevated tracts, while to
those of the plains it is as strange as an animal from a foreign
country. In my visits to the mountainous districts, I uniformly met with it, and, as far as the information of the natives
can be relied on, it is found on all the mountains *."

Now, if we were asked to conjecture how the Mydaus arrived
at the elevated regions of each of these isolated mountains, we
should say that before the isle was peopled by man, by whom
their numbers are now thinned, they may occasionally have
multiplied so as to be forced to collect together and migrate ; in
which case, notwithstanding the slowness of their motions,
some few would succeed in reaching another mountain, some
twenty, or even, perhaps, fifty miles distant : for although
the climate of the hot intervening plains would be unfavourable to them, they might support it for a time, and would find
there abundance of insects on which they feed. Volcanic eruptions, which at different times have covered the summits of
some of these lofty cones with steril sand and ashes, may have
occasionally contributed to force on these migrations.

The power of the terrestrial mammalia to cross the sea is
very limited, and we have already stated that the same species
is scarcely ever common to districts widely separated by the
ocean. If there be some exceptions to this rule they generally
admit of explanation, for there are natural means whereby some
animals may be floated across the water, and the sea sometimes
wears a passage through a neck of land, leaving individuals

* Zoological Researches in Java, No. 2.

of a species on each side of the new channel. Polar bears are known to have been frequently drifted on the ice from Greenland to Iceland; they can also swim to considerable distances, for Captain Parry, on the return of his ships through Barrow's Strait, met with a bear swimming in the water about midway between the shores, which were about forty miles apart, and where no ice was in sight *. " Near the east coast of Greenland," observes Scoresby, " they have been seen on the ice in such quantities, that they were compared to flocks of sheep on a common—and they are often found on field-ice, above two hundred miles from the shore †." Wolves, in the arctic regions, often venture upon the ice near the shore, for the purpose of preying upon young seals which they surprise when asleep. When these ice-floes get detached, the wolves are often carried out to sea, and though some may be drifted to islands or continents, the greater part of them perish, and have been often heard in this situation howling dreadfully, as they die by famine ‡.

During the short summer which visits Melville Island, various plants push forth their leaves and flowers the moment the snow is off the ground, and form a carpet spangled with the most lively colours. These secluded spots are reached annually by herds of musk-oxen and rein-deer, which travel immense distances over dreary and desolate regions, to graze undisturbed on these luxuriant pastures §. The rein-deer often pass along in the same manner, by the chain of the Aleutian Islands, from Behring's Straits to Kamtschatka, subsisting on the moss found in these islands during their passage ||.

Within the tropics there are no ice-floes; but, as if to compensate for that mode of transportation, there are floating isles of matted trees, which are often borne along through considerable spaces. These are sometimes seen sailing at the distance of fifty or one hundred miles from the mouth of the

* Append. to Parry's Second Voyage, years 1819–20.
† Account of the Arctic Regions, vol. i. p. 518.
‡ Turton, in a note to Goldsmith's Nat. Hist., vol. iii. p.43.
§ Supplement to Parry's First Voyage of Disc., p. 189.
|| Godman's American Nat. Hist., vol. i. p. 22.

H

Ganges, with living trees standing erect upon them. The Amazon, the Congo, and the Orinoco, also produce these verdant rafts, which are formed in the manner already described when speaking of the great raft of the Atchafalaya, an arm of the Mississippi, where a natural bridge of timber, ten miles long, and more than two hundred yards wide, has existed for more than forty years, supporting a luxurious vegetation, and rising and sinking with the water which flows beneath it *. That this enormous mass will one day break up and send down a multitude of floating islands to the gulf of Mexico, is the hope and well-founded expectation of the inhabitants of Louisiana.

On these green isles of the Mississippi, observes Malte-Brun†, young trees take root, and the pisliar and nenuphar display their yellow flowers ; there serpents, birds, and the cayman alligator, come to repose, and all are sometimes carried to the sea, and engulphed in its waters.

In a memoir lately published, a naval officer informs us, that as he returned from China by the eastern passage, he fell in, among the Moluccas, with several small floating islands of this kind, covered with mangrove-trees interwoven with underwood. The trees and shrubs retained their verdure, receiving nourishment from a stratum of soil which formed a white beach round the margin of each raft, where it was exposed to the washing of the waves and the rays of the sun ‡. The occurrence of soil in such situations may easily be explained, for all the natural bridges of timber which occasionally connect the islands of the Ganges, Mississippi, and other rivers, with their banks, are exposed to floods of water, densely charged with sediment.

Captain W. H. Smyth informs me, that when cruizing in the Cornwallis amidst the Philippine Islands, he has more than once seen, after those dreadful hurricanes called typhoons, floating islands of wood, with trees growing upon them, and that ships have sometimes been in imminent peril, in consequence of mistaking them for terra-firma.

---

* See ante, vol. i. p. 188.   † System of Geography, vol. v. p. 157.
‡ United Service Journal, No. 24, p. 697.

It is highly interesting to trace, in imagination, the effects of the passage of these rafts from the mouth of a large river to some archipelago, such as those in the South Pacific, raised from the deep, in comparatively modern times, by the operations of the volcano and the earthquake, and the joint labours of coral animals and testacea. If a storm arise, and the frail vessel be wrecked, still many a bird and insect may succeed in gaining, by flight, some island of the newly-formed group, while the seeds and berries of herbs and shrubs, which fall into the waves, may be thrown upon the strand. But if the surface of the deep be calm, and the rafts are carried along by a current, or wafted by some slight breath of air fanning the foliage of the green trees, it may arrive, after a passage of several weeks, at the bay of an island, into which its plants and animals may be poured out as from an ark, and thus a colony of several hundred new species may at once be naturalized.

We may remind the reader, that we merely advert to the transportation of these rafts as of extremely rare and accidental occurrence; but it may account, in tropical countries, for some of the rare exceptions to the general law, of the confined range of species.

Many of the cetacea, the whales of the northern seas for example, are found to desert one tract of the sea, and to visit another very distant, when they are urged by want of food or danger. The seals also retire from the coasts of Greenland in July, return again in September, and depart again in March, to return in June. They proceed in great droves northwards, directing their course where the sea is most free from ice, and are observed to be extremely fat when they set out on this expedition, and very lean when they come home again *.

Some naturalists have wondered that the sea-calves, dolphins, and other marine mammalia of the Mediterranean and Euxine should be identical with those found in the Caspian; and among other fanciful theories, they have suggested that they may dive through subterranean conduits, and thus pass from

---

* Krantz, vol. i. p. 129, cited by Goldsmith, Nat. Hist., vol. iii. p. 260.

one sea into the other. But as the occurrence of wolves and
other noxious animals, on both sides the British channel, was
adduced by Desmarest, as one of many arguments to prove that
England and France were once united, so the correspondence
of the aquatic species of the inland seas of Asia with those of
the Black Sea, tends to confirm the hypothesis for which there
are abundance of independent geological data, that those seas
were connected together by straits at no remote period of the
earth's history.

Geographical Distribution and Migrations of Birds.

We shall now offer a few observations on some of the other
divisions of the animal kingdom. Birds, notwithstanding their
great locomotive powers, form no exception to the general rules
already laid down, but, in this class as in plants and terrestrial
quadrupeds, different groups of species are circumscribed
within definite limits. We find, for example, one assemblage
in the Brazils, another in the same latitudes in central Africa,
another in India, and a fourth in New Holland. But some
species again are so local, that in the same archipelago, a single
island frequently contains a species found in no other spot on
the whole earth ; as is exemplified in some of the parrot tribes.
In this extensive family, which are, with few exceptions, inhabitants of tropical regions, the American group has not one
in common with the African, nor either of these with the parrots of India*.

Another illustration is afforded by that minute and beautiful
tribe, the humming-birds. The whole of them are, in the
first place, peculiar to the new world ; but there, although
some have a considerable range, as the *Trochilus flammifrons*
which is common to Lima, the island of Juan Fernandez and the
Straits of Magellan†, other species are peculiar to some of

* Prichard, vol. i. p. 47.

† Captain King, during his late survey, found this bird at the Straits of Magellan, in the month of May, the depth of winter, sucking the flowers of the large
species of fuchsia, then in bloom in the midst of a shower of snow.

the West-India islands, and have not been found elsewhere in the western hemisphere. The ornithology of our own country affords a no less striking exemplification of the same law, for the common grouse (*Tetrao scoticus*) occurs nowhere in the known world except in the British isles.

Some species of the vulture tribe are said to be true cosmopolites, and the common wild goose, *Anas anser*, Linn., if we may believe some ornithologists, is a general inhabitant of the globe, being met with from Lapland to the Cape of Good Hope, frequent in Arabia, Persia, China, and Japan, and in the American continent, from Hudson's Bay to South Carolina *. An extraordinary range has also been attributed to the nightingale, which extends from western Europe to Persia, and still farther. In a work entitled Specchio Comparativo †, by Charles Bonaparte, many species of birds are enumerated as common to Rome and Philadelphia; the greater part of these are migratory, but some of them, such as the long-eared owl, *Strix otus*, are permanent in both countries.

In parallel zones of the northern and southern hemispheres, a great general correspondence of form is observable, both in the aquatic and terrestrial birds, but there is rarely any specific identity; and this phenomenon is truly remarkable, when we recollect the readiness with which some birds, not gifted with great powers of flight, shift their quarters to different regions, and the facility with which others, possessing great strength of wing, perform their aërial voyages. Some migrate periodically from high latitudes, to avoid the cold of winter, and the accompaniments of cold,—scarcity of insects and vegetable food; others, it is said, for some particular kinds of nutriment required for rearing their young: for this purpose, they often traverse the ocean for thousands of miles, and re-cross it at other periods, with equal security.

Periodical migrations, no less regular, are mentioned by

* Bewick's Birds, vol. ii. p. 294, who cites Latham.
† Pisa, 1827 (not sold).

Humboldt, of many American water-fowl, from one part of the tropics to another in a zone where there is the same temperature throughout the year. Immense flights of ducks leave the valley of the Orinoco, when the increasing depth of its waters and the flooding of its shores prevent them from catching fish, insects, and aquatic worms. They then betake themselves to the Rio Negro and Amazon, having passed from the eighth and third degrees of north latitude, to the first and fourth of south latitude, directing their course south southeast. In September, when the Orinoco decreases and re-enters into its channel, these birds return northwards *.

The insectivorous swallows which visit our island would perish during winter, if they did not annually repair to warmer climes. It is supposed, that in these aërial excursions the average rapidity of their flight is not less than fifty miles an hour, so that when aided by the wind they soon reach warmer latitudes. Spallanzani calculated that the swallow can fly at the rate of ninety-two miles an hour, and conceived that the rapidity of the swift might be three times greater †. The rate of flight of the eider-duck (*Anas mollissima*) has been ascertained to be ninety miles an hour; and that of hawks and several other tribes, to be one hundred and fifty miles.

When we reflect how easily different species, in a great lapse of ages, may be each overtaken by gales and hurricanes, and, abandoning themselves to the tempest, be scattered at random through various regions of the earth's surface, where the temperature of the atmosphere, the vegetation, and the animal productions, might be suited to their wants, we shall be prepared to find some species capriciously distributed, and to be sometimes unable to determine the native countries of each. Captain Smyth informs me, that when engaged in his survey of the Mediterranean, he encountered a gale in the gulf of Lyons, at the distance of between twenty and thirty leagues from the coast of France, which bore along many land birds

---

* Voyage aux Régions Equinoxiales, tome vii. p. 429.

† Fleming, Phil. Zool., vol. ii. p. 43.

of various species, some of which alighted on the ship, while others were thrown with violence against the sails. In this manner islands become tenanted by species of birds inhabiting the nearest main land.

Geographical Distribution and Dissemination of Reptiles.

A few facts respecting the third great class of vertebrated animals, will suffice to show that the plan of nature, in regard to their location on the globe, is perfectly analogous to that already exemplified in other parts of the organic creation, and has probably been determined by similar causes.

Of the great saurians, the gavials which inhabit the Ganges differ from the cayman of America, or the crocodile of the Nile. The monitor of New Holland is specifically distinct from the Indian species ; these latter again from the African, and all from their congeners in the new world. So in regard to snakes ; we find the boa of America, represented by the python, a different though nearly allied genus, in India. America is the country of the rattle-snake, Africa of the cerastes, and Asia of the hooded snake or cobra di capello. There is a legend that St. Patrick expelled all reptiles from Ireland, and certain it is that none of the three species of snakes common in England, nor the toad, have been observed there by naturalists. They have our common frog, and our water-newt, and according to Ray (Quad. 264.) the green lizard (*Lacerta viridis*). Schultes the botanist observed, a few years since, in his tour in England, that there were two great islands in Europe of which the floras were unknown, Sardinia and Ireland ; we believe he might also have added the fauna of the latter country.

The range of the large reptiles is, in general, quite as limited as that of some orders of the terrestrial mammalia. The great saurians sometimes cross a considerable tract, in order to pass from one river to another ; but their motions by land are generally slower than those of quadrupeds. By water, however, they may transport themselves to distant situations more easily. The larger alligator of the Ganges sometimes descends beyond the brackish water of the Delta into the sea, and in such cases

*

it might chance to be drifted away by a current, and survive
till it reached a shore at some distance ; but such casualties are
probably very rare *.

Turtles migrate in large droves from one part of the ocean to
another during the ovipositing season. Dr. Fleming† mentions
that an individual of the hawk's bill turtle (*Chelonia imbricata*)
so common in the American seas has been taken at Papa Stour,
one of the West Zetland islands ; and according to Sibbald,
" the same animal came into Orkney." Another was taken in
1774, in the Severn, according to Turton. Two instances also of
the occurrence of the leathern tortoise (*C. coriacea*), on the coast
of Cornwall, in 1756, are mentioned by Borlase. These animals
of more southern seas can only be considered as stragglers
attracted to our shores during uncommonly warm seasons by an
abundant supply of food, or driven by storms to high latitudes.

Some of the smaller reptiles lay their eggs on aquatic plants ;
and these must often be borne rapidly by rivers, and conveyed
to distant regions, in a manner similar to the dispersion of
seeds before adverted to. But that the larger ophidians may be
themselves transported across the seas is evident from the following most interesting account of the arrival of one at the
island of St. Vincent. It is worthy of being recorded, says the
Rev. L. Guilding ‡, " that a noble specimen of the *Boa con-
strictor* was lately conveyed to us by the currents, twisted
round the trunk of a large sound cedar tree, which had probably been washed out of the bank by the floods of some great
South American river, while its huge folds hung on the
branches, as it waited for its prey. The monster was fortunately destroyed after killing a few sheep, and his skeleton
now hangs before me in my study, putting me in mind how
much reason I might have had to fear in my future rambles
through the forests of St. Vincent, had this formidable reptile
been a pregnant female, and escaped to a safe retreat."

---

* Malte-Brun says (Syst. of Geog. vol. viii. p. 193), that a crocodile is still preserved at Lyons that was taken from the *Rhone*, about two centuries ago; but no
particulars are given.   † Brit. Animals, p. 149; who cites Sibbald.
‡ Zool. Journ. vol. iii. p. 406. Dec. 1827.

CHAPTER VII.

Geographical Distribution and Migrations of Fish.

ALTHOUGH we are less acquainted with the habitations of marine animals than with the grouping of the terrestrial species before described, yet it is well ascertained that their distribution is governed by the same general laws. The testimony borne by MM. Péron and Lesueur to this important fact is remarkably strong. These eminent naturalists, after collecting and describing many thousand species which they brought to Europe from the southern hemisphere, insist most emphatically on their distinctness from those north of the equator ; and this remark they extend to animals of all classes, from those of a more simple to those of a more complex organization, from the sponges and medusæ to the cetacea. " Among all those which we have been able to examine," say they, " with our own eyes, or with regard to which it has appeared to us possible to pronounce with certainty, there is not a single animal of the southern regions which is not distinguished by essential characters from the analogous species in the northern seas *.

The fish of the Arabian gulf are said to differ entirely from

* Sur les Habitations des Animaux Marins. Ann. du Mus. tom. xv., cited by Prichard, Phys. Hist. of Mankind, vol. i. p. 51.

those of the Mediterranean, notwithstanding the proximity of these seas. The flying-fish are found (some stragglers excepted) only between the tropics,—in receding from the line they never approach a higher latitude than the fortieth parallel. Those inhabiting the Atlantic are said to be different species from those of the eastern ocean *. The electric gymnotus belongs exclusively to America, the trembler, or *Silurus electricus* to the rivers of Africa ; but the torpedo, or crampfish, is said to be dispersed over all tropical and many temperate seas †.

All are aware that there are certain fish of passage which have their periodical migrations like some tribes of birds. The salmon, towards the season of spawning, ascends the rivers for hundreds of miles, leaping up the cataracts which it meets in its course, and then retreats again into the depths of the ocean. The herring and the haddock, after frequenting certain shores in vast shoals for a series of years, desert them again and resort to other stations, followed by the species which prey on them. Eels are said to descend into the sea for the purpose of producing their young, which are seen returning into the freshwater by myriads, extremely small in size, but possessing the power of surmounting every obstacle which occurs in the course of a river, by applying their slimy and glutinous bodies to the surface of rocks, or the gates of a lock, even when dry, and so climbing over it ‡.

Gmelin says, that the anseres subsist in their migrations on the spawn of fish, and that oftentimes when they void the spawn, two or three days afterwards, the eggs retain their vitality unimpaired §. When there are many disconnected freshwaterlakes in a mountainous region, at various elevations, each remote from the other, it has often been deemed inconceivable how they could all become stocked with fish from one common source ; but it has been suggested, that the minute eggs of

* Malte-Brun, vol. i. p. 507.   † Ibid.
‡ Phil. Trans. 1747, p. 395.   § Amœn, Acad., Essay 75.

these animals may sometimes be entangled in the feathers of water-fowl. These, when they alight to wash and plume themselves in the water, may often unconsciously contribute to propagate swarms of fish, which, in due season, will supply them with food. Some of the water-beetles, also, as the dyticidæ, are amphibious, and in the evening quit their lakes and pools, and flying in the air transport the minute ova of fishes to distant waters. In this manner some naturalists account for the fry of fish appearing occasionally in small pools caused by heavy rains.

Geographical Distribution and Migrations of Testacea.

The testacea, of which so great a variety of species occurs in the sea, are a class of animals of peculiar importance to the geologist, because their remains are found in strata of all ages, and generally in a higher state of preservation than those of other organic beings. Climate has a decided influence on the geographical distribution of species in this class ; but as there is much greater uniformity of temperature in the waters of the ocean, than in the atmosphere which invests the land, the diffusion of many marine molluscs is extensive.

Some forms, as those of the nautili, volutæ, and cyprææ, attain their fullest development in warm latitudes ; and most of their species are exclusively confined to them. Péron and Lesueur remark, that the Haliotis gigantea, of Van Dieman's land, and the Phasianella, diminish in size as they follow the coasts of New Holland to King George's straits, and entirely disappear beyond them *. Almost all the species of South American shells differ from those of the Indian archipelago in the same latitudes ; and on the shores of many of the isles of the South Pacific, peculiar species have been obtained. But we are as yet by no means able to sketch out the submarine provinces of shells, as the botanist has done those of the terrestrial, and even of the subaqueous plants. There can be little

* Ann. du Mus. d'Hist. Nat., tom. xv.

doubt, however, that the boundaries in this case, both of latitude and longitude, will be found in general well defined. The continuous lines of continents, stretching from north to south, prevent a particular species from belting the globe, and following the direction of the isothermal lines. The inhabitants of the West Indian seas, for example, cannot enter the Pacific without passing round through the inclement climate of Cape Horn.

Currents also flowing permanently in certain directions, and the influx at certain points of great bodies of fresh-water, limit the extension of many species. Those which love deep water are arrested by shoals ; others, fitted for shallow seas, cannot migrate across unfathomable abysses. Some few species, however, have an immense range, as the *Bulla aperta* for example, which is found in almost all zones. The habitation of the *Bulla striata* extends from the shores of Egypt to the coasts of England and France, and it recurs again in the seas of Senegal, Brazil, and the West Indies. The *Turbo petræus* inhabits the seas of England, Guadaloupe, and the Cape of Good Hope *, and many instances of a similar kind might be enumerated. The *Ianthina fragilis* has wandered into almost every sea both tropical and temperate. This " common oceanic snail" derives its buoyancy from an admirably contrived float, which has enabled it not only to disperse itself so universally, but to become an active agent in disseminating other species which attach themselves, or their ova, to its shell †.

It is evident that among the testacea, as in plants and the higher orders of animals, there are species which have a power of enduring a wide range of temperature, whereas others cannot resist a considerable change of climate. Among the fresh-

* Fér. Art. Geogr. Phys. Dict. Class. d'Hist. Nat.

† Mr. Broderip possesses specimens of *Ianthina fragilis*, bearing more than one species of barnacle (*Pentelasmis*), presented to him by Captain King and Lieutenant Graves. One of these specimens, taken alive by Captain King far at sea, and a little north of the equator, is so loaded with those cirrhipeds, and with numerous ova, that all the upper part of its shell is invisible.

water molluscs, and those which breathe air, Férrussac mentions a few instances of species of almost universal diffusion.

The Helix putris (*Succinea putris*, Lam.) so common in Europe, where it reaches from Norway to Italy, is also found in Egypt, in the United States, in Newfoundland, Jamaica, Tranquebar, and, it is even said, in the Marianne Isles. As this animal inhabits constantly the borders of pools and streams where there is much moisture, it is not impossible that different water-fowl have been the agents of spreading some of its minute eggs, which may have been entangled in their feathers. *Helix aspersa*, one of the commonest of our larger land-shells, is found in South America, at the foot of Chimborazo, as also in Cayenne. Some conchologists have conjectured, that it was accidentally imported in some ship; for it is an eatable species, and these animals are capable of retaining life, during long voyages, without air or nourishment *.

Mr. Lowe, in a memoir just published in the Cambridge Transactions, enumerates seventy-one species of land mollusca, collected by him in the islands of Madeira and Porto Santo, sixty of which belonged to the genus Helix alone, including as subgenera Bulimus and Achatina, and excluding Vitrina and Clausilia;—forty-four of these are new. It is remarkable, that very few of the above-mentioned species are common to the neighbouring archipelago of the Canaries; but it is a still more striking fact, that of the sixty species of the three genera above-mentioned, thirty-one are natives of Porto Santo; whereas, in Madeira, which contains ten times the superficies, were found but twenty-nine. Of these only four were common

---

* Four individuals of a large species of Bulimus, from Valparaiso, were brought to England by Lieutenant Graves, who accompanied Captain King in his late expedition to the Straits of Magellan. They had been packed up in a box and enveloped in cotton, two for a space of thirteen, one for seventeen, and a fourth for upwards of twenty months; but on being exposed, by Mr. Broderip, to the warmth of a fire in London, and provided with tepid water and leaves, they revived, and are now living in Mr. Loddiges' palm-house.

to the two islands, which are only separated by a distance of twelve leagues; and two even of these four (namely, *Helix rhodostoma* and *H. ventrosa*) are species of general diffusion, common to Madeira, the Canaries, and the South of Europe*.

The confined range of these molluscs may easily be explained, if we admit that species have only one birth-place; and the only problem to be solved would relate to the exceptions—to account for the dissemination of some species throughout several isles and the European continent. May not the eggs, when washed into the sea by the undermining of cliffs, float uninjured to a distant shore ?

Notwithstanding the proverbially slow motion of snails and molluscs in general, and although many aquatic species adhere constantly to the same rock for their whole lives, they are by no means destitute of provision for disseminating themselves rapidly over a wide area. Some lay their eggs in a sponge-like nidus, wherein the young remain enveloped for a time after their birth, and this buoyant substance floats far and wide as readily as sea-weed. The young of other viviparous tribes are often borne along, entangled in sea-weed. Sometimes they are so light that, like grains of sand, they can be easily moved by currents. Balani and serpulæ are sometimes found adhering to floating cocoa-nuts, and even to fragments of pumice. In rivers and lakes, on the other hand, aquatic univalves usually attach their eggs to leaves and sticks which have fallen into the water, and which are liable to be swept away, during floods, from tributaries to the main streams, and from thence to all parts of the same basins. Particular species may thus migrate during one season from the head waters of the Mississippi, or any other great river, to countries bordering the sea, at the distance of many thousand miles.

An illustration of the mode of attachment of these eggs will be seen in the annexed diagram. (No. 1.)

* Camb. Phil. Trans. vol. iv. 1831.

(No. 1.)

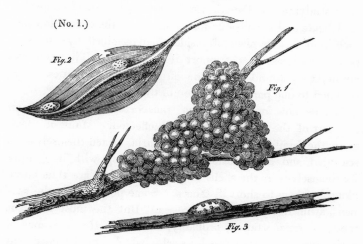

Eggs of fresh-water Molluscs.

Fig. 1. Eggs of Ampullaria ovata (a fluviatile species), fixed to a small sprig which had fallen into the water.

Fig. 2. Eggs of Planorbis albus, attached to a dead leaf lying under water.

Fig. 3. Eggs of the common Limneus (L. vulgaris), adhering to a dry stick under water.

The habit of some testacea to adhere to floating wood is proved by their fixing themselves to the bottoms of ships. By this mode of conveyance *Mytilus polymorphus* has been brought from northern Europe to the Commercial Docks in the Thames, where the species is now domiciled.

Geographical Distribution and Migrations of Zoophytes.

Zoophytes are very imperfectly known, but there can be little doubt that each maritime region possesses species peculiar to itself. The madrepores, or lamelliferous polyparia, are found in their fullest development only in the tropical seas of Polynesia and the East and West Indies, and this family is represented only by a few species in our seas. Those even of the Mediterranean are inferior in size, and, for the most part, different from such as inhabit the tropics. Péron and Lesueur, after studying the Holothuriæ, Medusæ, and other congeners of delicate and changeable forms, came to the conclusion that each kind has its place of residence determined by the temperature necessary to support its existence. Thus, for example, they found the abode of Pyrosoma Atlantica to be confined to one particular region of the Atlantic ocean *.

Let us now inquire how the transportation of polyps from one part of the globe to another is effected. Many of them, as in the families Flustra and Sertularia, attach themselves to sea-weed, and are occasionally drifted along with it. Many fix themselves to the shells of gasteropods, and are thus borne along by them to short distances. Some polyps, like the sea-pens, swim freely about in the sea. But the most frequent mode of transportation, probably, consists in the buoyancy of their eggs, or certain small vesicles which are detached and are capable of becoming the foundation of a new colony. These gems, as they have been called, may be swept along by a wave that breaks upon a coral-reef, and may then be borne by a current to a distance.

That some zoophytes adhere to floating bodies is proved by their being found attached to the bottoms of ships, as in the case of testacea before alluded to.

Geographical Distribution and Migrations of Insects.

Before we conclude our sketch of the manner in which the habitable parts of the earth are shared out among particular assemblages of organic beings, we must offer a few remarks on insects, which, by their numbers, and the variety of their powers and instincts, exert a prodigious influence in the economy of animate nature. As a large portion of these minute creatures are strictly dependent for their subsistence on certain species of vegetables, the entomological provinces must coincide in a considerable degree with the botanical.

All the insects, says Latreille, brought from the eastern parts

---

* Voy. aux Terres Australes, tome i. p. 492.

of Asia and China, whatever be their latitude and temperature, are distinct from those of Europe and of Africa. The
insects of the United States, although often they approach
very close to our own, are nevertheless specifically distinguishable by some characters. In South America, the equinoctial
lands of New Grenada and Peru on the one side, and of Guiana
on the other, contain for the most part distinct groups; the
Andes forming the division, and interposing a narrow line
of severe cold between climates otherwise very similar *.

The insects of the United States, even those of the northern
provinces as far as Canada, differ specifically from the European, while those of Greenland appear to be in a great measure
identical with our own. Some insects are very local, while
a few, on the contrary, are common to remote countries, between which the torrid zone and the ocean intervene. Thus
our painted lady butterfly (*Vanessa Cardui*) reappears in New
Holland and Japan with scarcely a varying streak †. The
same species is said to be one of the few insects which are
universally dispersed over the earth, being found in Europe,
Asia, Africa, and America; and its wide range is the more
interesting because it seems explained by its migratory instinct,
seconded, no doubt, by a capacity enjoyed by few species, of
enduring a great diversity of temperature.

A vast swarm of this species, forming a column from ten to
fifteen feet broad, was, a few years since, observed in the
Canton de Vaud; they traversed the country with great
rapidity from north to south, all flying onwards in regular
order, close together, and not turning from their course on
the approach of other objects. Professor Bonelli, of Turin,
observed, in March of the same year, a similar swarm of the
same species, also directing their flight from north to south, in
Piedmont, in such immense numbers, that at night the flowers
were literally covered with them. They had been traced from

* Géographie Générale des Insectes et des Arachnides. Mém. du Mus. d'Hist
Nat. tome iii.

† Kirby and Spence, vol. iv. p. 487.

Coni, Raconi, Susa, &c.   A similar flight at the end of the last century is recorded by M. Louch, in the Memoirs of the Academy of Turin.   The fact is the more worthy of notice, because the caterpillars of this butterfly are not gregarious, but solitary from the moment that they are hatched; and this instinct remains dormant, while generation after generation passes away, till it suddenly displays itself in full energy when their numbers happen to be in excess.

Not only peculiar species but certain types distinguish particular countries; and there are groups, observes Kirby, which represent each other in distant regions, whether in their form, their functions, or in both.   Thus the honey and wax of Europe, Asia, and Africa, are in each case prepared by bees congenerous with our common hive-bee (*Apis*, Latr.); while in America, this genus is nowhere indigenous, but is replaced by Melipona and Trigona; and in New Holland by a still different, but undescribed type *.

As almost all insects are winged, they can readily spread themselves wherever their progress is not opposed by un-congenial climates, or by seas, mountains, and other phy-sical impediments; and these barriers they can sometimes sur-mount by abandoning themselves to violent winds, which, as we before stated, when speaking of floating seeds, may in a few hours carry them to very considerable distances.   On the Andes some sphinxes and flies have been observed by Hum-boldt, at the height of nineteen thousand one hundred and eighty feet above the sea, and which appeared to him to have been involuntarily carried into these regions by ascending currents of air †.

White mentions a remarkable shower of aphides which seem to have emigrated, with an east wind, from the great hop plantations of Kent and Sussex, and blackened the shrubs and vegetables where they alighted at Selbourne, spreading at the same time in great clouds all along the vale from Farnham to

* Kirby and Spence, vol. iv. p. 497.
† Description of the Equatorial Regions—Malte-Brun, vol. v. p. 379.

Alton.   These aphides are sometimes accompanied by vast numbers of the common lady-bird (*Coccinella semptem-punctata*), which feed upon them *.

It is remarkable, says Kirby, that many of the insects which are occasionally observed to emigrate, as, for instance, the libellulæ, coccinellæ, carabi, cicadæ, &c., are not usually social insects, but seem to congregate, like swallows, merely for the purpose of emigration †.   Here, therefore, we have an example of an instinct developing itself on certain rare emergencies, causing unsocial species to become gregarious, and to venture sometimes even to cross the ocean.

To the armies of locusts darkening the air in Africa, and traversing the globe from Turkey to our southern counties in England, we need not here allude.   When the western gales sweep over the Pampas, they bear along with them myriads of insects of various kinds.   As a proof of the manner in which species may be thus diffused, we may mention that when the Creole frigate was lying in the outer roads off Buenos Ayres, in 1819, at the distance of six miles from the land, her decks and rigging were suddenly covered with thousands of flies and grains of sand.   The sides of the vessel had just received a fresh coat of paint, to which the insects adhered in such numbers as to spot and disfigure the vessel, and to render it necessary partially to renew the paint ‡.   Captain W. H. Smyth was obliged to repaint his vessel, the Adventure, in the Mediterranean, from the same cause.   He was on his way from Malta to Tripoli, when a southern wind blowing from the coast of Africa, then one hundred miles distant, drove such myriads of flies upon the fresh paint, that not the smallest point was left unoccupied by insects.

To the southward of the river Plate, off Cape St. Antonio, and at the distance of fifty miles from land, several large dragon-flies alighted on the Adventure frigate, during Captain King's late expedition to the Straits of Magellan.   If

---

* Kirby and Spence, vol. ii. p. 9, 1817.   † Vol. ii. p. 12.
    ‡ I am indebted to Lieutenant Graves, R.N. for this information.

the wind abates when insects are thus crossing the sea, the most delicate species are not necessarily drowned, for many can repose without sinking on the unruffled surface of the deep. The slender, long-legged tipulæ have been seen standing on the surface of the sea, when driven out far from our coast, and took wing immediately on being approached *. Exotic beetles are sometimes thrown on our shore, which revive after having been long drenched in salt-water; and the periodical appearance of some conspicuous butterflies amongst us, after being unseen for five or fifty years, has been ascribed, not without probability, to the agency of the winds.

Inundations of rivers, observes Kirby, if they happen at any season except in the depth of winter, always carry down a number of insects, floating on the surface of bits of stick, weeds, &c., so that when the waters subside, the entomologist may generally reap a plentiful harvest. In the dissemination, moreover, of these minute beings, as in that of plants, the larger animals play their part. Insects are, in numberless instances, borne along in the coats of animals, or the feathers of birds; and the eggs of some species are capable, like seeds, of resisting the digestive powers of the stomach, and after they are swallowed with herbage, may be ejected again unharmed in the dung.

Geographical Distribution and Diffusion of Man.

We have reserved for the last our observations on the range and diffusion of the human species over the earth, and the influence of man, in spreading other animals and plants, especially the terrestrial.

Many naturalists have amused themselves in speculating on the probable birth-place of mankind, the point from which, if we assume the whole human race to have descended from a single pair, the tide of emigration must originally have proceeded. It has been always a favourite conjecture, that this birth-place was situated within or near the tropics, where per-

* I state this fact on the authority of my friend, Mr. Curtis.

petual summer reigns, and where fruits, herbs, and roots, are plentifully supplied throughout the year. The climate of these regions, it has been said, is suited to a being born without any covering, and who had not yet acquired the arts of building habitations or providing clothes.

" The hunter state," it has been argued, " which Montesquieu placed the first, was probably only the second stage to which mankind arrived, since so many arts must have been invented to catch a salmon or a deer, that society could no longer have been in its infancy when they came into use *." When regions where the spontaneous fruits of the earth abound became overpeopled, men would naturally diffuse themselves over the neighbouring parts of the temperate zone; but a considerable time would probably elapse before this event took place; and it is possible, as a writer before cited observes, that in the interval before the multiplication of their numbers and their increasing wants had compelled them to emigrate, some arts to take animals were invented, but far inferior to what we see practised at this day among savages. As their habitations gradually advanced into the temperate zone, the new difficulties they had to encounter would call forth by degrees the spirit of invention, and the probability of such inventions always rises with the number of people involved in the same necessity †.

A distinguished modern writer, who coincides for the most part in the views of Aphonin above mentioned, has introduced one of the persons in his second dialogue as objecting to the theory of the human race having gradually advanced from a savage to a civilized state, on the ground that " the first man must have inevitably been destroyed by the elements or devoured by savage beasts, so infinitely his superiors in physical force ‡." He then contends against the difficulty here started

---

* Rev. J. F. Brand, commenting on Aphonin, Amœn. Acad. vol. vii. p. 409. Brand's Select Dissert. from the Amœn. Acad. vol. i. p. 118.

† Idem., ib.

‡ Sir H. Davy, Consolations in Travel, p. 74.

by various arguments, all of which were, perhaps, superfluous, for if a philosopher is pleased to indulge in conjectures on this subject, why should he not assign, as the original seat of man, some one of those large islands within the tropics, which are as free from wild beasts as Van Diemen's Land or Australia? Here man may have remained for a period peculiar to a single isle, just as some of the large anthropomorphous species are now limited to one island within the tropics. In such a situation, the new-born race might have lived in security, though far more helpless than the New Holland savages, and might have found abundance of vegetable food. Colonies may afterwards have been sent forth from this mother country, and then the peopling of the earth may have proceeded according to the hypothesis before alluded to.

In an early stage of society the necessity of hunting acts as a principle of repulsion, causing men to spread with the greatest rapidity over a country, until the whole is covered with scattered settlements. It has been calculated that eight hundred acres of hunting-ground only produce as much food as half an acre of arable land. When the game has been in a great measure exhausted, and a state of pasturage succeeds, the several hunter tribes, being already scattered, may multiply in a short time into the greatest number which the pastoral state is capable of sustaining. The necessity, says Brand, thus imposed upon the two savage states, of dispersing themselves far and wide over the country, affords a reason why, at a very early period, the worst parts of the earth may have become inhabited.

But this reason it may be said is only applicable in as far as regards the peopling of a continuous continent; whereas the smallest islands, however remote from continents, have almost invariably been found inhabited by man. St. Helena, it is true, afforded an exception; for when that island was discovered in 1501, it was only inhabited by sea-fowl, and occasionally by seals and turtles, and was covered with a forest of trees and shrubs, all of species peculiar, as we before observed, with one or two exceptions, and which seem to have been expressly created for this remote and insulated spot.

But very few of the numerous coral islets and volcanos of the vast Pacific, capable of sustaining a few families of men, have been found untenanted, and we have, therefore, to inquire whence and by what means, if all the members of the great human family have had one common source, could those savages have migrated. Cook, Forster, and others have remarked that parties of savages in their canoes must often have lost their way and must have been driven on distant shores, where they were forced to remain, deprived both of the means and of the requisite intelligence for returning to their own country. Thus Captain Cook found on the island Wateoo, three inhabitants of Otaheite, who had been drifted thither in a canoe, although the distance between the two isles is five hundred and fifty miles. In 1696, two canoes containing thirty persons, who had left Ancorso, were thrown by contrary winds and storms on the island of Samar, one of the Philippines, at a distance of eight hundred miles. In 1721, two canoes, one of which contained twenty-four, and the other six persons, men, women, and children, were drifted from an island called Baroilep, to the island of Guam, one of the Marians *.

Kotzebue, when investigating the Coral isles of Radack, at the eastern extremity of the Caroline isles, became acquainted with a person of the name of Kadu, who was a native of Ulea, an isle fifteen hundred miles distant, from which he had been drifted with a party. Kadu and three of his countrymen, one day, left Ulea in a sailing boat, when a violent storm arose, and drove them out of their course ; they drifted about the open sea for eight months, according to their reckoning by the moon, making a knot on a cord at every new moon. Being expert fishermen they subsisted entirely on the produce of the sea; and when the rain fell, laid in as much fresh-water as they had vessels to contain it. " Kadu," says Kotzebue, " who was the

---

* Malte-Brun's Geography, vol. iii. p. 419.

best diver, frequently went down to the bottom of the sea, where it is well known that the water is not so salt, with a cocoa nut shell, with only a small opening." When these unfortunate men reached the isles of Radack, every hope and almost every feeling had died within them; their sail had long been destroyed, their canoe had long been the sport of winds and waves, and they were picked up by the inhabitants of Aur, in a state of insensibility *; but by the hospitable care of those islanders they soon recovered, and were restored to perfect health.

Captain Beechey, in his late voyage to the Pacific, fell in with some natives of the Coral Islands, who had in a similar manner been carried to a great distance from their native country. They had embarked to the number of one hundred and fifty souls, in three double canoes, from Anaa, or Chain Island, situate about three hundred miles to the eastward of Otaheite. They were overtaken by the monsoon, which dispersed the canoes, and after driving them about the ocean, left them becalmed, so that a great number of persons perished. Two of the canoes were never heard of, but the other was drifted from one uninhabited island to another, at each of which the voyagers obtained a few provisions; and at length, after having wandered for a distance of six hundred miles, they were found and carried to their home in the Blossom †.

The space traversed in some of these instances was so great, that similar accidents might suffice to transport canoes from various parts of Africa to the shores of South America, or from Spain to the Azores, and thence to North America. So that man, even in a rude state of society, is liable to be scattered involuntarily by the winds and waves over the globe, in a manner singularly analogous to that in which many plants and animals are diffused. We ought not then to wonder that during the ages required for some tribes of the human race to attain that advanced stage of civilization which empowers the

* Kotzebue's Voyage, 1815—1818. Quarterly Review, vol. xxvi. p. 361.
† Narrative of a Voyage to the Pacific, &c., in the years 1825, 1826, 1827, 1828, p. 170.

navigator to cross the ocean in all directions with security, the whole earth should have become the abode of rude tribes of hunters and fishers. Were the whole of mankind now cut off, with the exception of one family, inhabiting the old or new continent, or Australia, or even some coral islet of the Pacific, we should expect their descendants, though they should never become more enlightened than the South Sea Islanders or the Esquimaux, to spread in the course of ages over the whole earth, diffused partly by the tendency of population to increase beyond the means of subsistence, in a limited district, and partly by the accidental drifting of canoes by tides and currents to distant shores.

Involuntary influence of Man in diffusing Animals and Plants.

Many of the general remarks which we made respecting the influence of man in spreading or in checking the diffusion of plants, apply equally to his relations with the animal kingdom. We shall be led on a future occasion to speak of the instrumentality of our species in naturalizing useful animals and plants in new regions, when we explain our views of the effects which the spreading and increase of certain species exert in the extirpation of others. At present we shall confine ourselves to a few remarks on the involuntary aid which man lends to the dissemination of species.

In the mammiferous class our influence is chiefly displayed in increasing the number of quadrupeds which are serviceable to us, and in exterminating or reducing the number of those which are noxious.

Sometimes, however, we unintentionally promote the multiplication of inimical species, as when we introduced the rat, which was not indigenous in the New World, into all parts of America. They have been conveyed over in ships, and now infest a great multitude of islands and parts of that continent. In like manner the Norway rat has been imported into England, where it plunders our property in ships and houses.

The great viper, fer de lance, a species no less venomous than the rattle-snake, which now ravages Martinique and St. Lucia, was accidentally introduced by man, and exists in no other part of the West Indies.

Many parasitic insects, which attack our persons, and some of which are supposed to be peculiar to our species, have been carried into all parts of the earth, and have as high a claim as man to an *universal* geographical distribution.

A great variety of insects have been transported in ships from one country to another, especially in warmer latitudes. Notwithstanding the coldness of our climate, we have been unable to prevent the cockroach (*Blatta orientalis*), from entering and diffusing itself in our ovens and kneading troughs, and availing itself of the artificial warmth which we afford. It is well known also that beetles, and many other kinds of ligniperdous insects, have been introduced into Great Britain in timber; especially several North American species. " The commercial relations," says Malte-Brun *, " between France and India, have transported from the latter country the aphis, which destroys the apple-tree, and two sorts of Neuroptera, the *lucifuga* and *flavicola,* mostly confined to Provence and the neighbourhood of Bordeaux, where they devour the timber in the houses and naval arsenals."

Among molluscs we may mention the *teredo navalis*, which is a native of equatorial seas, but which, by adhering to the bottom of ships, was transported to Holland, where it has been most destructive to vessels and piles. The same species has also become naturalized in England, and other countries enjoying an extensive commerce.

In all these and innumerable other instances, we may regard the involuntary agency of man as strictly analogous to that of the inferior animals. Like them we unconsciously contribute to extend or limit the geographical range and numbers of certain species, in obedience to general rules in the economy of nature, which are for the most part beyond our control.

* Syst. of Geog., vol. viii., p. 169.

CHAPTER VIII.

It would be superfluous to examine the various attempts which were made to explain the phenomena of the distribution of species alluded to in the preceding chapters, in the infancy of the sciences of botany, zoology, and physical geography. The theories or rather conjectures then indulged, now stand refuted by a simple statement of facts; and if Linnæus were living, he would be the first to renounce the notions which he promulgated. For he imagined the habitable world to have been for a certain time limited to one small tract, the only portion of the earth's surface that was as yet laid bare by the subsidence of the primæval ocean. In this fertile spot he supposed the originals of all the species of plants which exist on this globe to have been congregated, together with the first ancestors of all animals and of the human race. "In quâ commodè habitaverint animalia omnia, et vegetabilia lætè germinaverint." In order to accommodate the various habitudes of so many creatures, and to provide a diversity of climate suited to their several natures, the tract in which the creation took place was supposed to have been situated in some warm region of the earth, but to have contained a lofty mountain range, on the heights and in the declivities of which were to be found

*

all temperatures and every clime, from the torrid to the frozen zone *.

That there never was a universal ocean since the planet was inhabited, or rather since the oldest groups of strata yet known to contain organic remains were formed, is proved by the presence of terrestrial plants in all the older formations ; and if this conclusion was not established, yet no geologist could deny that since the first small portion of the earth was laid dry, there have been many entire changes in the species of plants and animals inhabiting the land.

But without dwelling on the above and other refuted theories, let us inquire whether we can substitute some hypothesis as simple as that of Linnæus, to which the phenomena now ascertained in regard to the distribution both of aquatic and terrestrial species may be referred. The following may, perhaps, be reconcileable with known facts:—Each species may have had its origin in a single pair, or individual, where an individual was sufficient, and species may have been created in succession at such times and in such places as to enable them to multiply and endure for an appointed period, and occupy an appointed space on the globe.

In order to explain this theory, let us suppose every living thing to be destroyed in the western hemisphere, both on the land and in the ocean, and permission to be given to man to people this great desert, by transporting into it animals and plants from the eastern hemisphere, a strict prohibition being enforced against introducing two original stocks of the same species.

Now the result we conceive of such a mode of colonizing would correspond exactly, so far as regards the grouping of animals and plants, with that now observed throughout the globe. It would be necessary for naturalists, before they imported species into particular localities, to study attentively the climate and other physical conditions of each spot. It would

* De terra habitabili incremento; also Prichard, Phys. Hist. of Mankind, vol. i. p. 17, where the hypotheses of different naturalists are enumerated.

be no less requisite to introduce the different species in succession, so that each plant and animal might have time and
opportunity to multiply before the species destined to prey
upon it was admitted.   Many herbs and shrubs, for example,
must spread far and wide before the sheep, the deer, and the
goat could be allowed to enter, lest they should devour and
annihilate the original stocks of many plants, and then perish
themselves for want of food.   The above-mentioned herbivorous animals in their turn must be permitted to make
considerable progress before the entrance of the first pair of
wolves or lions.   Insects must be allowed to swarm before the
swallow could be permitted to skim through the air and feast
on thousands at one repast.

It is evident that, however equally in this case our original
stocks were distributed over the whole surface of land and
water, there would nevertheless arise distinct botanical and
zoological provinces, for there are a great many natural barriers
which oppose common obstacles to the advance of a variety
of species.   Thus, for example, almost all the animals and
plants naturalized by us towards the extremity of South
America, would be unable to spread beyond a certain limit,
towards the east, west, and south, because they would be
stopped by the ocean, and a few of them only would succeed in
reaching the cooler latitudes of the northern hemisphere, because they would be incapable of bearing the heat of the
tropics, through which they must pass.   In the course of ages,
undoubtedly, exceptions would arise, and some species might
become common to the temperate and polar regions, or both
sides of the equator ; for we have before shown that the
powers of diffusion conferred on some classes are very great.
But we should confidently predict that these exceptions would
never become so numerous as to invalidate the general rule.

Some of the plants and animals transplanted by us to the
coast of Chili or Peru would never be able to cross the Andes,
so as to reach the Eastern plains ; nor, for a similar reason,
would those first established in the Pampas, or the valleys of

the Amazon and the Orinoco, ever arrive at the shores of the Pacific.

In the ocean an analogous state of things would prevail; for there, also, climate would exert a great influence in limiting the range of species, and the land would stop the migrations of aquatic tribes as effectually as the sea arrests the dispersion of the terrestrial. As certain birds, insects, and the seeds of plants, can never cross the direction of prevailing winds, so currents form natural barriers to the dissemination of many oceanic races. A line of shoals may be as impassable to pelagian species, as are the Alps and the Andes to plants and animals peculiar to plains, while deep abysses may prove insuperable obstacles to the migrations of the inhabitants of shallow waters.

It is worthy of observation, that one effect of the introduction of single pairs of each species must be the confined range of certain groups in spots which, like small islands, or solitary inland lakes, have few means of interchanging their inhabitants with adjoining regions. Now this congregating, in a small space, of many peculiar species, would give an appearance of *centres* or *foci* of creation, as they have been termed, as if there were favourite points where the creative energy has been in greater action than in others, and where the numbers of peculiar organic beings have consequently become more considerable.

We do not mean to call in question the soundness of the inferences of some botanists, as to the former existence of certain limited spots whence species of plants have been propagated, radiating, as it were, in all directions from a common centre. On the contrary, we conceive these phenomena to be the necessary consequences of the plan of nature before suggested, operating during the successive mutations of the surface, some of which the geologist can prove to have taken place subsequently to the period when many species now existing were created. In order to exemplify how this arrangement of plants may have been produced, let us imagine that, about three centuries before the discovery of St. Helena (itself of submarine volcanic origin), a multitude of new isles had been thrown up in the surrounding sea, and that these had each become clothed with plants emigrating from St. Helena, in the same manner as the wild plants of Campania have diffused themselves over Monte Nuovo. Whenever the first botanist investigated the new archipelago, he would, in all probability, find a different assemblage of plants in each of the isles of recent formation; but in St. Helena itself, he would meet with individuals of every species belonging to all parts of the archipelago, and some, in addition, peculiar to itself, viz., those which had not been able to obtain a passage into any one of the surrounding new-formed lands. In this case, it might be truly said that the original isle was the primitive focus, or centre, of a certain type of vegetation, whereas, in the surrounding isles, there would be a smaller number of species, yet all belonging to the same group.

But this peculiar distribution of plants would not warrant the conclusion that, in the space occupied by St. Helena, there had been a greater exertion of creative power than in the spaces of equal area occupied by the new adjacent lands, because, within the period in which St. Helena had acquired its peculiar vegetation, each of the spots supposed to be subsequently converted into land, may have been the birth-places of a great number of *marine* animals and plants, which may have had time to scatter themselves far and wide over the southern Atlantic.

Perhaps it may be objected to some part of the foregoing train of reasoning, that during the lapse of past ages, especially during many partial revolutions of the globe of comparatively modern date, different zoological and botanical provinces ought to have become more confounded and blended together—that the distribution of species approaches too nearly to what might have been expected, if animals and plants had been introduced into the globe when its physical geography had already assumed the features which it now wears; whereas we know that, in certain districts, considerable geographical changes have taken place since species identical with those now in being were created.

These, and many kindred topics, cannot be fully discussed until we have considered, not merely the general laws which may regulate the first introduction of species, but those which may limit their *duration* on the earth. Brocchi, whose untimely death in Egypt is deplored by all who have the progress of geology at heart, has remarked, when hazarding some interesting conjectures respecting " the loss of species," that a modern naturalist had no small assurance, who declared " that individuals alone were capable of destruction, and that species were so perpetuated that nature could not annihilate them, so long as the planet lasted, or at least that nothing less than the shock of a comet, or some similar disaster, could put an end to their existence*." The Italian geologist, on the contrary, had satisfied himself, that many species of testacea, which formerly inhabited the Mediterranean, had become extinct, although a great number of others, which had been the contemporaries of those lost races, still survived. He came to the opinion, that about half the species which peopled the waters when the Subapennine strata were deposited, had gone out of existence ; and in this inference he does not appear to have been far wrong.

But instead of seeking a solution of this problem, like some other geologists of his time, in a violent and general catastrophe, Brocchi endeavoured to imagine some regular and constant law by which species might be made to disappear from the earth gradually and in succession. The death, he suggested, of a species might depend, like that of individuals, on certain peculiarities of constitution conferred upon them at their birth, and as the longevity of the one depends on a certain force of vitality, which, after a period, grows weaker and weaker, so the duration of the other may be governed by the quantity of prolific power bestowed upon the species, which, after a season, may decline in energy, so that the fecundity and multiplication of

* Necker, Phytozool. Philosoph., p. 21. Brocchi, Conch. Foss. Subap., tome i., p. 229.

individuals may be gradually lessened from century to century, " until that fatal term arrives, when the embryo, incapable of extending and developing itself, abandons, almost at the instant of its formation, the slender principle of life by which it was scarcely animated,—and so all dies with it."

Now we might coincide in opinion with the Italian naturalist, as to the gradual extinction of species one after another, by the operation of regular and constant causes, without admitting an inherent principle of deterioration in their physiological attributes. We might concede " that many species are on the decline, and that the day is not far distant when they will cease to exist;" yet deem it consistent with what we know of the nature of organic beings, to believe that the last individuals of each species retain their prolific powers in their full intensity.

Brocchi has himself speculated on the share which a change of climate may have had in rendering the Mediterranean unfit for the habitation of certain testacea, which still continued to thrive in the Indian ocean, and of others which were now only represented by analogous forms within the tropics. He must also have been aware that other extrinsic causes, such as the progress of human population, or the increase of some one of the inferior animals, might gradually lead to the extirpation of a particular species, although its fecundity might remain to the last unimpaired. If, therefore, amid the vicissitudes of the animate and inanimate world, there are known causes capable of bringing about the decline and extirpation of species, it became him thoroughly to investigate the full extent to which these might operate, before he speculated on any cause of so purely hypothetical a kind, as " the diminution of the pro-lific virtue."

If it could have been shown that some wild plant had insensibly dwindled away and died out, as sometimes happens to cultivated varieties propagated by cuttings, even though climate, soil, and every other circumstance should continue identically the same—if any animal had perished while the physical condition of the earth, and the number and force of its foes, with every other extrinsic cause, remained unaltered, then might we have some ground for suspecting that the infirmities of age creep on as naturally on species as upon individuals. But in the absence of such observations, let us turn to another class of facts, and examine attentively the circumstances which determine the *stations* of particular animals and plants, and perhaps we shall discover, in the vicissitudes to which these stations are exposed, a cause fully adequate to explain the phenomena under consideration.

Stations comprehend all the circumstances, whether relating to the animate or inanimate world, which determine whether a given plant or animal can exist in a given locality, so that if it be shown that stations can become essentially modified by the influence of known causes, it will follow that species, as well as individuals, are mortal.

Every naturalist is familiar with the fact, that although in a particular country, such as Great Britain, there may be more than three thousand species of plants, ten thousand insects, and a great variety in each of the other classes, yet there will not be more than a hundred, perhaps not half that number, inhabiting any given locality. There may be no want of space in the supposed tract; it may be a large mountain, or an extensive moor, or a great river-plain, containing room enough for individuals of every species in our island; yet the spot will be occupied by a few to the exclusion of many, and these few are enabled, throughout long periods, to maintain their ground successfully against every intruder, notwithstanding the facilities which species enjoy, by virtue of their powers of diffusion, of invading adjacent territories.

The principal causes which enable a certain assemblage of plants thus to maintain their ground against all others depend, as is well known, on the relations between the physiological nature of each species, and the climate, exposure, soil, and other physical conditions of the locality. Some plants live only on rocks, others in meadows, a third class in marshes. Of the latter, some delight in a fresh-water morass,—others in salt marshes, where their roots may copiously absorb saline particles. Some prefer an alpine region in a warm latitude, where, during the heat of summer, they are constantly irrigated by the cool waters of melting snows. To others loose sand, so fatal to the generality of species, affords the most proper station. The *Carex arenaria* and the *Elymus arenarius* acquire their full vigour on a sandy dune, obtaining an ascendency over the very plants which in a stiff clay would immediately stifle them.

Where the soil of a district is of so peculiar a nature that it is extremely favourable to certain species, and agrees ill with every other, the former get exclusive possession of the ground, and, as in the case of heaths, live in societies. In like manner, the Bog moss (*Hypnum palustre*) is fully developed in peaty swamps, and becomes, like the heath, in the language of botanists, a social plant. Such monopolies would be very frequent, if the powers of a great number of species were not equally balanced, and if animals did not interfere most actively to preserve an equilibrium in the vegetable kingdom.

" All the plants of a given country," says Decandolle in his usual spirited style, " are at war one with another. The first which establish themselves by chance in a particular spot, tend, by the mere occupancy of space, to exclude other species—the greater choke the smaller, the longest livers replace those which last for a shorter period, the more prolific gradually make themselves masters of the ground, which species multiplying more slowly would otherwise fill."

In this continual strife, it is not always the resources of the plant itself which enable it to maintain or extend its ground. Its success depends, in a great measure, on the number of its foes or allies among the animals and plants inhabiting the same region. Thus, for example, a herb which loves the shade may multiply, if some tree with spreading boughs and dense foliage flourish in the neighbourhood. Another, which, if unassisted, would be overpowered by the rank growth of some hardy competitor, is secure, because its leaves are unpalatable to cattle, which, on the other hand, annually crop down its antagonist, and rarely suffer it to ripen its seed.

Oftentimes we see some herb which has flowered in the midst of a thorny shrub, when all the other individuals of the same species, in the sunny fields around, are eaten down, and cannot bring their seed to maturity. In this case, the shrub has lent his armour of spines and prickles to protect the defence-less herb against the mouths of the cattle, and thus a few individuals which occupied, perhaps, the most unfavourable station in regard to exposure, soil, and other circumstances, may nevertheless, by the aid of an ally, become the principal source whereby the winds are supplied with seeds which per-petuate the species throughout the surrounding tract.

In the above example we see one plant shielding another from the attacks of animals; but instances are, perhaps, still more numerous, where some animal defends a plant against the enmity of some other subject of the vegetable kingdom.

Scarcely any beast, observes a Swedish naturalist *, will touch the nettle, but fifty different kinds of insects are fed by it. Some of these seize upon the root, others upon the stem ; some eat the leaves, others devour the seeds and flowers : but for this multitude of enemies, the nettle would annihilate a great number of plants. Linnæus tells us, in his Tour in Scania, that goats were turned into an island which abounded with the *Agrostis arundinacea*, where they perished by famine ; but horses, which followed them, grew fat on the same plant. The goat, also, he says, thrives on the meadow-sweet and water hemlock, plants which are injurious to cattle †.

Every plant, observes Wilcke, has its proper insect allotted to it to curb its luxuriancy, and to prevent it from multiplying to the exclusion of others. " Thus grass in meadows sometimes flourishes so as to exclude all other plants : here the Phalæna graminis (*Bombyx gram.*), with her numerous progeny, find a well-spread table ; they multiply in immense numbers, and the

* Wilcke, Amœn. Acad., vol. vi., p. 17, § 12.   † Ibid., vol. vii., 409.

farmer for some years laments the failure of his hay crop ; but the grass being consumed, the moths die with hunger, or remove to another place. Now the quantity of grass being greatly diminished, the other plants, which were before choked by it, spring up, and the ground becomes variegated with a multitude of different species of flowers. Had not nature given a commission to this minister for that purpose, the grass would destroy a great number of species of vegetables, of which the equilibrium is now kept up *."

In the above passage allusion is made to the ravages committed in 1740, and the two following years, in many provinces of Sweden, by a most destructive insect. The same moth is said never to touch the fox-tail grass †, so that it may be classed as a most active ally and benefactor of that species, and as peculiarly instrumental in preserving it in its present abundance. A discovery of Rolander, cited in the treatise of Wilcke above-mentioned, affords a good illustration of the checks and counterchecks which nature has appointed to preserve the balance of power amongst species. " The Phalæna strobilella has the fir cone assigned to it to deposit its eggs upon ; the young caterpillars coming out of the shell consume the cone and superfluous seed ; but lest the destruction should be too general, the Ichneumon strobilellæ lays its eggs in the caterpillar, inserting its long tail in the openings of the cone till it touches the included insect, for its body is too large to enter. Thus it fixes its minute egg upon the caterpillar, which being hatched destroys it ‡."

Entomologists enumerate many parallel cases where insects, appropriated to certain plants, are kept down by other insects, and these again by parasites expressly appointed to prey on them §. Few, perhaps, are in the habit of duly appreciating the extent to which insects are active in preserving the balance

---

* Wilcke, Amœn. Acad., vol. vi., p. 17, § 11 and 12.
† Kirby and Spence, vol. i., p. 178.
‡ Wilcke, ibid., § 14.   § Kirby and Spence, vol. iv., p. 218.

of species among plants, and thus regulating indirectly the relative numbers of many of the higher orders of terrestrial animals.

The peculiarity of their agency consists in their power of suddenly multiplying their numbers, to a degree which could only be accomplished in a considerable lapse of time in any of the larger animals, and then as instantaneously relapsing, without the intervention of any violent disturbing cause, into their former insignificance.

If for the sake of employing, on different but rare occasions, a power of many hundred horses, we were under the necessity of feeding all these animals at great cost in the intervals when their services were not required, we should greatly admire the invention of a machine, such as the steam-engine, which was capable, at any moment, of exerting the same degree of strength without any consumption of food during periods of inaction. The same kind of admiration is strongly excited when we contemplate the powers of insect life, in the creation of which nature has been so prodigal. A scanty number of minute individuals, only to be detected by careful research, are ready in a few days, weeks, or months, to give birth to myriads which may repress any degree of monopoly in another species, or remove nuisances, such as dead carcasses, which might taint the air. But no sooner has the destroying commission been executed, than the gigantic power becomes dormant—each of the mighty host soon reaches the term of its transient existence, and the season arrives when the whole species passes naturally into the egg, and thence into the larva and pupa state. In this defenceless condition it may be destroyed either by the elements, or by the augmentation of some of its numerous foes which may prey upon it in these stages of its transformation ; or it often happens that, in the following year, the season proves unfavourable to the hatching of the eggs or the development of the pupæ.

Thus the swarming myriads depart which may have covered the vegetation like the aphides, or darkened the air like locusts.

In almost every season there are some species which in this manner put forth their strength, and then, like Milton's spirits which thronged the spacious hall, " reduce to smallest forms their shapes immense"—

—————— So thick the aëry crowd
Swarm'd and were straiten'd ; till, the signal given,
Behold a wonder ! they but now who seem'd
In bigness to surpass earth's giant sons,
Now less than smallest dwarfs.

A few examples will illustrate the mode in which this force operates. It is well known that among the countless species of the insect creation, some feed on animal, others on vegetable matter, and, upon considering a catalogue of eight thousand British insects and arachnidæ, Mr. Kirby found that these two divisions were nearly a counterpoise to each other, the carnivorous being somewhat preponderant. There are also distinct species, some appointed to consume living, others dead or putrid animal and vegetable substances. One female, of *Musca carnaria,* will give birth to twenty thousand young ; and the larvæ of many flesh-flies devour so much food in twenty-four hours, and grow so quickly, as to increase their weight two hundredfold ! In five days after being hatched they arrive at their full growth and size, so that there was ground, says Kirby, for the assertion of Linnæus, that three flies of M. vomitoria could devour a dead horse as quickly as a lion * ; and another Swedish naturalist remarks, that so great are the powers of propagation of a single species, even of the smallest insects, that each can commit, when required, more ravages than the elephant †.

Next to locusts, the aphides, perhaps, exert the greatest power over the vegetable world, and, like them, are sometimes so numerous as to darken the air. The multiplication of these little creatures is without parallel, and almost every plant has its peculiar species. Reaumur has proved, that in five generations one aphis may be the progenitor of 5,904,900,000 descendants ; and it is supposed that in one year there may be

———————————

* Kirby and Spence, vol. i., p. 250.
† Wilcke, Amœn. Acad., chap. ii.

twenty generations *.   Mr. Curtis † observes, that as among
caterpillars we find some that are constantly and unalterably
attached to one or more particular species of plants, and others
that feed indiscriminately on most sorts of herbage, so it is precisely with the aphides ; some are particular, others more
general feeders; and as they resemble other insects in this
respect, so they do also in being more abundant in some years
than others.   In 1793 they were the chief, and in 1798 the
sole cause of the failure of the hops.   In 1794, a season almost
unparalleled for drought, the hop was perfectly free from them,
while peas and beans, especially the former, suffered very
much from their depredations.

The ravages of the caterpillars of some of our smaller moths
afford a good illustration of the temporary increase of a species.
The oak-trees of a considerable wood have been stripped of
their leaves as bare as in winter, by the caterpillars of a small
green moth (*Tortrix irridana*,) which has been observed the
year following not to abound ‡.   The Gamma moth (*Plusia
gamma*), although one of our common species, is not dreaded
by us for its devastations, but legions of their caterpillars have,
at times, created alarm in France, as in 1735.   Reaumur observes, that the female moth lays about four hundred eggs ; so
that if twenty caterpillars were distributed in a garden, and all
lived through the winter and became moths in the succeeding
May, the eggs laid by these, if all fertile, would produce eight
hundred thousand §.   A modern writer, therefore, justly observes, that did not Providence put causes in operation to keep
them in due bounds, the caterpillars of this moth alone, leaving
out of consideration the two thousand other British species,
would soon destroy more than half of our vegetation ||.

In the latter part of the last century an ant, most destructive

* Kirby and Spence, vol. i., p. 174.

† Trans. Linn. Soc., vol. vi.

‡ Lib. Ent. Know., Insect Trans., p. 203.   See Haworth Lep.

§ Reaumur, ii. 237.

|| Lib. Ent. Know., Insect Trans., p. 212.

to the sugar-cane (*Formica saccharivora*), appeared in such infinite hosts, in the island of Grenada, as to put a stop to the cultivation of that vegetable.   Their numbers were incredible. The plantations and roads were filled with them ; many domestic quadrupeds, together with rats, mice, and reptiles, and even birds, perished in consequence of this plague.   It was not till 1780 that they were at length annihilated by torrents of rain, which accompanied a dreadful hurricane *.

We may conclude by mentioning some instances of the devastations of locusts in various countries.   Among other parts of Africa, Cyrenaica has been at different periods infested by myriads of these creatures, which have consumed nearly every green thing.   The effect of the havoc committed by them may be estimated by the famine they occasioned.   St. Augustin mentions a plague of this kind in Africa which destroyed no less than eight hundred thousand men in the kingdom of Masanissa alone, and many more upon the territories bordering upon the sea.   It is also related, that in the year 591 an infinite army of locusts migrated from Africa into Italy, and, after grievously ravaging the country, were cast into the sea, when there arose a pestilence from their stench which carried off nearly a million of men and beasts.

In the Venetian territory also, in 1478, more than thirty thousand persons are said to have perished in a famine, occasioned by this scourge ; and other instances are recorded of their devastations in France, Spain, Italy, Germany, &c.   In different parts of Russia also, Hungary, and Poland,—in Arabia and India, and other countries, their visitations have been periodically experienced.   Although they have a preference for certain plants, yet, when these are consumed, they will attack almost all the remainder.   In the accounts of the invasions of locusts, the statements which appear most marvellous relate to the prodigious mass of matter which encumbers the sea wherever they are blown into it, and the pestilence arising from its putrefaction.   Their dead bodies are said to have been,

* Kirby and Spence, vol. i., p. 183.   Castle, Phil. Trans., xxx., 346.

in some places, heaped one upon another, to the depth of four feet, in Russia, Poland, and Lithuania ; and when in southern Africa they were driven into the sea by a north-west wind, they formed, says Barrow, along the shore, for fifty miles, a bank three or four feet high *. But when we consider that forests are stripped of their foliage, and the earth of its green garment, for thousands of square miles, it may well be supposed that the volume of animal matter produced may equal that of great herds of quadrupeds and flights of large birds suddenly precipitated into the sea.

The occurrence of such events at certain intervals, in hot countries, like the severe winters and damp summers returning after a series of years in the temperate zone, affect the proportional numbers of almost all classes of animals and plants, and are probably fatal to the existence of many which would otherwise thrive there, while, on the contrary, they must be favourable to certain species which, if deprived of such aid, might not maintain their ground.

Although it may usually be remarked that the extraordinary increase of some one species is immediately followed and checked by the multiplication of another, yet this is not always the case, partly because many species feed in common on the same kinds of food, and partly because many kinds of food are often consumed indifferently by one and the same species. In the former case, where a variety of different animals have precisely the same taste, as, for example, when many insectivorous birds and reptiles devour alike some particular fly or beetle, the unusual numbers of the latter may only cause a slight and almost imperceptible augmentation of each of those species of bird and reptile. In the other instance, where one animal preys on others of almost every class, as, for example, where our English buzzards devour not only small quadrupeds, as rabbits and field-mice, but also birds, frogs, lizards, and insects, the profusion of any one of these last may cause all such general feeders to subsist more exclusively upon

* Travels in Africa, p. 257. Kirby and Spence, vol. i., p. 215.

the species thus in excess, and the balance may thus be restored.

The number of species which are nearly omnivorous is considerable; and although every animal has, perhaps, a predilection for some one description of food rather than another, yet some are not even confined to one of the great kingdoms of the organic world. Thus when the racoon of the West Indies can neither procure fowls, fish, snails, nor insects, it will attack the sugar-canes, and devour various kinds of grain. The civets, when animal food is scarce, maintain themselves on fruits and roots.

Numerous birds, which feed indiscriminately on insects and plants, are perhaps more instrumental than any other of the terrestrial tribes in preserving a constant equilibrium between the relative numbers of different classes of animals and vegetables. If the insects become very numerous and devour the plants, these birds will immediately derive a larger portion of their subsistence from insects, just as the Arabians, Syrians, and Hottentots feed on locusts, when the locusts devour their crops.

The intimate relation of the inhabitants of the water to those of the land, and the influence exerted by each on the relative number of species, must not be overlooked amongst the complicated causes which determine the existence of animals and plants in certain regions. A large proportion of the amphibious quadrupeds and reptiles prey partly on aquatic plants and animals, and in part on terrestrial; and a deficiency of one kind of prey causes them to have immediate recourse to the other. The voracity of certain insects, as the dragon-fly, for example, is confined to the water during one stage of their transformations, and in their perfect state to the air. Innumerable water-birds both of rivers and seas derive in like manner their food indifferently from either element; so that the abundance or scarcity of prey in one induces them either to forsake or more constantly to haunt the other. Thus an intimate connexion between the state of the animate creation in a lake or river, and in the adjoining dry land, is maintained; or between a continent, with its lakes and rivers, and the ocean. It is well known that many birds migrate, during stormy seasons, from the sea-shore into the interior, in search of food; while others, on the contrary, urged by like wants, forsake their inland haunts, and live on substances rejected by the tide.

The migrations of fish into rivers during the spawning season supplies another link of the same kind. Suppose the salmon to be reduced in numbers by some marine foes, as by seals and grampuses, the consequence must often be, that in the course of a few years the otters at the distance of several hundred miles inland will be lessened in number from the scarcity of fish. On the other hand, if there be a dearth of food for the young fry of the salmon in rivers and estuaries, so that few return to the sea, the sand-eels and other marine species, which are usually kept down by the salmon, will swarm in greater profusion.

It is unnecessary to accumulate a greater number of illustrations in order to prove that the stations of different plants and animals depend on a great complication of circumstances,— on an immense variety of relations in the state of the animate and inanimate worlds. Every plant requires a certain climate, soil, and other conditions, and often the aid of many animals, in order to maintain its ground. Many animals feed on certain plants, being often restricted to a small number, and sometimes to one only; other members of the animal kingdom feed on plant-eating species, and thus become dependent on the conditions of the *stations* not only of their prey, but of the plants consumed by them.

Having duly reflected on the nature and extent of these mutual relations in the different parts of the organic and inorganic worlds, we may next proceed to examine the results which may be anticipated from the fluctuations now continually in progress in the state of the earth's surface, and in the geographical distribution of its living productions.

CHAPTER IX.

The circumstances which constitute the *Stations* of Animals are changeable—Extension of the range of one species alters the condition of others—Supposed effects which may [have followed the first entrance of the Polar Bears into Iceland—The first appearance of a new species in a region causes the chief disturbance—Changes known to have resulted from the advance of human population—Whether man increases the productive powers of the earth—Indigenous Quadrupeds and Birds of Great Britain known to have been extirpated—Extinction of the Dodo—Rapid propagation of the domestic Quadrupeds over the American Continent—Power of exterminating species no prerogative of Man—Concluding Remarks.

WE have seen that the stations of animals and plants depend not merely on the influence of external agents in the inanimate world, and the relations of that influence to the structure and habits of each species, but also on the state of the contemporary living beings which inhabit the same part of the globe. In other words, the possibility of the existence of a certain species in a given locality, or of its thriving more or less therein, is determined not merely by temperature, humidity, soil, elevation, and other circumstances of the like kind, but also by the existence or non-existence, the abundance or scarcity, of a particular assemblage of other plants and animals in the same region.

If we show that both these classes of circumstances, whether relating to the animate or inanimate creation, are perpetually changing, it will follow that species are subject to incessant vicissitudes; and if the result of these mutations, in the course of ages, be so great as materially to affect the general condition of *stations*, it will follow that the successive destruction of species must now be part of the regular and constant order of Nature.

It will be desirable, first, to consider the effects which every extension of the numbers or geographical range of one species must produce on the condition of others inhabiting the same regions. When the necessary consequences of such extensions have been fully explained, the reader will be prepared to appreciate the important influence which slight modifications in the physical geography of the globe may exert on the condition of organic beings.

In the first place it is clear, that when any region is stocked with as great a variety of animals and plants as the productive powers of that region will enable it to support, the addition of any new species, or the *permanent* numerical increase of one previously established, must always be attended either by the local extermination or the numerical decrease of some other species.

There may undoubtedly be considerable fluctuations from year to year, and the equilibrium may be again restored without any permanent alteration; for in particular seasons a greater supply of heat, humidity, or other causes may augment the total quantity of vegetable produce, in which case all the animals subsisting on vegetable food, and others which prey on them, may multiply without any one species giving way; but whenever the aggregate quantity of vegetable produce remains unaltered, the progressive increase of one animal or plant implies the decline of another.

All agriculturists and gardeners are familiar with the fact, that when weeds intrude themselves into the space appropriated to cultivated species, the latter are starved in their growth or stifled. If we abandon for a short time a field or garden, a host of indigenous plants,

The darnel, hemlock, and rank fumitory,

pour in and obtain the mastery, extirpating the exotics, or putting an end to the monopoly of some native plants.

If we inclose a park, and stock it with as many deer as the herbage will support, we cannot add sheep without lessening the number of the deer; nor can other herbivorous species be subsequently introduced, unless the individuals of each species in the park become fewer in proportion.

So if there be an island where leopards are the only beasts of prey, and the lion, tiger, and hyæna afterwards enter, the leopards, if they stand their ground, will be reduced in number. If the locusts then arrive and swarm greatly, it may deprive a large number of phytophagous animals of their food, and thereby cause a famine, not only among them, but among the beasts of prey;—certain species, perhaps, which had the weakest footing in the island will thus be annihilated.

We have seen how many distinct geographical provinces there are of aquatic and terrestrial species, and how great are the powers of migration conferred on different classes, whereby the inhabitants of one region may be enabled from time to time to invade another, and do actually so migrate and diffuse themselves over new countries. Now, although our knowledge of the history of the animate creation dates from so recent a period, that we can scarcely trace the advance or decline of any animal or plant, except in those cases where the influence of man has intervened, yet we can easily conceive what must happen when some new colony of wild animals or plants enters a region for the first time, and succeeds in establishing itself.

Let us consider how great are the devastations committed at certain periods by the Greenland bears, when they are drifted to the shores of Iceland in considerable numbers on the ice. These periodical invasions are formidable even to man ; so that when the bears arrive, the inhabitants collect together, and go in pursuit of them with fire-arms—each native who slays one being rewarded by the king of Denmark. The Danes of old, when they landed in their marauding expeditions upon our coast, hardly excited more alarm ; nor did our islanders muster more promptly for the defence of their lives and property against a common enemy, than the modern Icelanders against these formidable brutes. It frequently happens, says Henderson, that the natives are pursued by the bear when he has been long at sea, and when his natural ferocity has been strengthened by the keenness of hunger; if unarmed, it is frequently by stratagem only that they make their escape *.

Let us cast our thoughts back to the period when the first polar bears reached Iceland, before it was colonized by the Norwegians in 874;—we may imagine the breaking up of an immense barrier of ice, like that which, in 1816 and the following year, disappeared from the east coast of Greenland, which it had surrounded for four centuries. By the aid of such means of transportation, a great number of these quadrupeds might effect a landing at the same time, and the havoc which they would make among the species previously settled in the island would be terrific. The deer, foxes, seals, and even birds, on which these animals sometimes prey, would be soon thinned down.

But this would be a part only, and probably an insignificant portion, of the aggregate amount of change brought about by the new invader. The plants on which the deer fed being less consumed in consequence of the lessened numbers of that herbivorous species, would soon supply more food to several insects, and probably to some terrestrial testacea, so that the latter would gain ground. The increase of these would furnish other insects and birds with food, so that the numbers of these last would be augmented. The diminution of the seals would afford a respite to some fish which they had persecuted; and these fish, in their turn, would then multiply and press upon their peculiar prey. Many water-fowls, the eggs and young of which are devoured by foxes, would increase when the foxes were thinned down by the bears ; and the fish on which the water-fowls subsisted would then, in their turn, be less numerous. Thus the numerical proportions of a great number of the inhabitants, both of the land and sea, might be permanently altered by the settling of one new species in the region ; and the changes caused indirectly might ramify through all classes of the living creation, and be almost endless.

* Journal of a Residence in Iceland, p. 276.

An actual illustration of what we have here only proposed hypothetically, is in some degree afforded by the selection of small islands by the eider duck for its residence during the season of incubation ; its nests being seldom, if ever, found on the shores of the main land, or even of a large island. The Icelanders are so well aware of this, that they have expended a great deal of labour in forming artificial islands, by separating from the main-land certain promontories, joined to it by narrow isthmuses. This insular position is necessary to guard against the destruction of the eggs and young birds, by foxes, dogs, and other animals. One year, says Hooker *, it happened that, in the small island of Vidoc, adjoining the coast of Iceland, a fox got over *upon the ice,* and caused great alarm, as an immense number of ducks were then sitting on their eggs or young ones. It was long before he was taken, which was at last, however, effected by bringing another fox to the island, and fastening it by a string near the haunt of the former, by which he was allured within shot of the hunter.

It is usually the first appearance of an animal or plant, in a region to which it was previously a stranger, that gives rise to the chief alteration ; since, after a time, an equilibrium is again established. But it must require ages before such a new adjustment of the relative forces of so many conflicting agents can be definitively settled. The causes in simultaneous action are so numerous, that they admit of an almost infinite number of combinations ; and it is necessary that all these should have occurred once before the total amount of change, capable of flowing from any new disturbing force, can be estimated.

Thus, for example, suppose that once in two centuries a frost of unusual intensity, or a volcanic eruption of immense violence, accompanied by floods from the melting of glaciers, should occur in Iceland ; or an epidemic disease, fatal to the larger number of individuals of some one species, and not affecting others,—these, and a variety of other contingencies, all of which

* Tour in Iceland, vol. i., p. 64, second edition.

L

may occur at once, or at periods separated by different intervals of time, ought to happen before it would be possible for us to declare what ultimate alteration the presence of any new comer, such as the bear before mentioned, might occasion in the animal population of the isle.

Every new condition in the state of the organic or inorganic creation, a new animal or plant, an additional snow-clad mountain, any permanent change, however slight in comparison to the whole, gives rise to a new order of things, and may make a material change in regard to some one or more species. Yet a swarm of locusts, or a frost of extreme intensity, may pass away without any great apparent derangement; no species may be lost, and all may soon recover their former relative numbers, because the same scourges may have visited the region, again and again, at some former periods. Every plant that was incapable of resisting such a degree of cold, every animal which was exposed to be entirely cut off by famine, in consequence of the consumption of vegetation by the locusts, may have perished already, so that the subsequent recurrence of similar catastrophes is attended only by a temporary change.

We are best acquainted with the mutations brought about by the progress of human population, and the growth of plants and animals favoured by man. To these, therefore, we should, in the first instance, turn our attention. If we conclude, from the concurrent testimony of history and of the evidence yielded by geological data, that man is, comparatively speaking, of very modern origin, we must at once perceive how great a revolution in the state of the animate world the increase of the human race, considered merely as consumers of a certain quantity of organic matter, must necessarily cause.

It may, perhaps, be said, that man has, in some degree, compensated for the appropriation to himself of so much food, by artificially improving the natural productiveness of soils, by irrigation, manure, and a judicious intermixture of mineral ingredients conveyed from different localities. But it admits of reasonable doubt, whether, upon the whole, we fertilize or impoverish the lands which we occupy. This assertion may seem startling to many, because they are so much in the habit of regarding the sterility or productiveness of land in relation to the wants of man, and not as regards the organic world generally. It is difficult, at first, to conceive, if a morass is converted into arable land, and made to yield a crop of grain, even of moderate abundance, that we have not improved the capabilities of the habitable surface—that we have not empowered it to support a larger quantity of organic life. In such cases, a tract, before of no utility to man, may be reclaimed and become of high agricultural importance, but it may yield, at the same time, a scantier vegetation. If a lake be drained and turned into a meadow, the space will provide sustenance to man and many terrestrial animals serviceable to him, but not perhaps so much food as it previously yielded to the aquatic races.

If the pestiferous Pontine Marshes were drained and covered with corn, like the plains of the Po, they might, perhaps, feed a smaller number of animals than they do now; for these morasses are filled with of herds of buffaloes and swine, and they swarm with birds, reptiles, and insects.

The felling of dense and lofty forests which covered, even within the records of history, a considerable space on the globe, now tenanted by civilized man, must usually have lessened the amount of vegetable food throughout the space where these woods grew. We must also take into our account the area covered by towns, and a still larger surface occupied by roads.

If we force the soil to bear extraordinary crops one year, we are, perhaps, compelled to let it lie fallow the next. But nothing so much counterbalances the fertilizing effects of human art as the extensive cultivation of foreign herbs and shrubs, which, although they are often more nutritious to man, seldom thrive with the same rank luxuriance as the native plants of a district. Man is, in truth, continually striving to diminish

L 2

the natural diversity of the *stations* of animals and plants in every country, and to reduce them all to a small number fitted for species of economical use. He may succeed perfectly in attaining his object, even though the vegetation be comparatively meagre, and the total amount of animal life be greatly lessened.

Spix and Martius have given a lively description of the incredible number of insects which lay waste the crops in Brazil, besides swarms of monkeys, flocks of parrots and other birds, as well as the paca, agouti, and wild swine. They describe the torment which the planter and the naturalist suffer from the musquitoes, and the devastation of the ants and blattæ ; they speak of the dangers to which they were exposed from the jaguar, the poisonous serpents, lizards, scorpions, centipedes, and spiders. But with the increasing population and cultivation of the country, observe these naturalists, these evils will gradually diminish ; when the inhabitants have cut down the woods, drained the marshes, made roads in all directions, and founded villages and towns, man will by degrees triumph over the rank vegetation and the noxious animals, and all the elements will second and amply recompense his activity *.

The number of human beings now peopling the earth is supposed to amount to eight hundred millions, so that we may easily understand how great a number of beasts of prey, birds, and animals of every class, this prodigious population must have displaced, independently of the still more important consequences which have followed from the derangement brought about by man in the relative numerical strength of particular species.

Let us make some inquiries into the extent of the influence which the progress of society has exerted, during the last seven or eight centuries, in altering the distribution of our indigenous British animals. Dr. Fleming has prosecuted this inquiry with his usual zeal and ability, and in a memoir † on the subject

* Travels in Brazil, vol. i., p. 260.
† Ed. Phil. Journ., No. xxii., p. 287.   Oct. 1824.

has enumerated the best-authenticated examples of the decrease or extirpation of certain species during a period when our population has made the most rapid advances. We shall offer a brief outline of his results.

The stag, as well as the fallow deer and the roe, were formerly so abundant that, according to Lesley, from five hundred to a thousand were sometimes slain at a hunting-match; but the native races would already have been extinguished, had they not been carefully preserved in certain forests. The otter, the marten, and the polecat, were also in sufficient numbers to be pursued for the sake of their fur ; but they have now been reduced within very narrow bounds. The wild cat and fox have also been sacrificed throughout the greater part of the country, for the security of the poultry-yard or the fold. Badgers have been expelled from nearly every district which at former periods they inhabited.

Besides these, which have been driven out from some haunts, and everywhere reduced in number, there are some which have been wholly extirpated ; such as the ancient breed of indigenous horses, the wild boar, and the wild oxen, of which last, however, a few remains are still preserved in the parks of some of our nobility. The beaver, which was eagerly sought after for its fur, had become scarce at the close of the ninth century, and, by the twelfth century, was only to be met with, according to Giraldus de Barri, in one river in Wales, and another in Scotland. The wolf, once so much dreaded by our ancestors, is said to have maintained its ground in Ireland so late as the beginning of the eighteenth century (1710), though it had been extirpated in Scotland thirty years before, and in England at a much earlier period. The bear, which in Wales was regarded as a beast of the chace equal to the hare or the boar *, only perished as a native of Scotland in the year 1057 †:

Many native birds of prey have also been the subjects of unremitting persecution. The eagles, larger hawks, and ravens, have disappeared from the more cultivated districts. The

---

* Ray, Syn. Quad., p. 214.   † Ibid., p 295.

haunts of the mallard, the snipe, the redshank, and the bittern, have been drained equally with the summer dwellings of the lapwing and the curlew. But these species still linger in some portion of the British isles; whereas the large capercailzies, or wood grouse, formerly natives of the pine-forests of Ireland and Scotland, have been destroyed within the last fifty years. The egret and the crane, which appear to have been formerly very common in Scotland, are now only occasional visitants *.

The bustard (*Otis tarda*), observes Graves in his British Ornithology †, " was formerly seen in the downs and heaths of various parts of our island, in flocks of forty or fifty birds; whereas it is now a circumstance of rare occurrence to meet with a single individual." Bewick also remarks, " that they were formerly more common in this island than at present; they are now found only in the open counties of the south and east, in the plains of Wiltshire, Dorsetshire, and some parts of Yorkshire." In the few years that have elapsed since Bewick wrote, this bird has entirely disappeared from Wiltshire and Dorsetshire ‡.

These changes, we may observe, are derived from very imperfect memorials, and relate only to the larger and more conspicuous animals inhabiting a small spot on the globe; but they cannot fail to exalt our conception of the enormous revolutions which, in the course of several thousand years, the whole human species must have effected.

The kangaroo and the emu are retreating rapidly before the progress of colonization in Australia; and it scarcely admits of doubt, that the general cultivation of that country must lead to the extirpation of both. The most striking example of the loss, even within the last two centuries, of a remarkable species, is that of the dodo—a bird first seen by the Dutch when they landed on the Isle of France, at that time uninhabited, immediately after the discovery of the passage to the East Indies by

* Fleming, Syn. Quad., p. 295.   † Vol. iii. London, 1821.
‡ Land Birds, vol. i., p. 316, Ed. 1821.

the Cape of Good Hope. It was of a large size and singular
form ; its wings short, like those of an ostrich, and wholly
incapable of sustaining its heavy body even for a short flight:
In its general appearance it differed from the ostrich, casso-
wary, or any known bird.

Many naturalists gave figures of the dodo after the commencement of the seventeenth century, and there is a painting
of it in the British Museum, which is said to have been taken
from a living individual. Beneath the painting is a leg, in a
fine state of preservation, which ornithologists are agreed cannot
belong to any other known bird. In the museum at Oxford,
also, there is a foot and a head, in an imperfect state, but M.
Cuvier doubts the identity of this species with that of which
the painting is preserved in London *.

In spite of the most active search, during the last century,
no information respecting the dodo was obtained, and some
authors have gone so far as to pretend that it never existed ;
but amongst a great mass of satisfactory evidence in favour
of the recent existence of this species, we may mention that an
assemblage of fossil bones were recently discovered, under
a bed of lava, in the Isle of France, and sent to the Paris
museum by M. Desjardins. They almost all belonged to a
large living species of land-tortoise, called *Testudo Indica*, but
amongst them were the head, sternum, and humerus of the
dodo. M. Cuvier showed me these valuable remains in Paris,
and assured me that they left no doubt in his mind that the
huge bird was one of the gallinaceous tribe †.

---

* Some have complained that inscriptions on tomb-stones convey no general
information except that individuals were born and died, accidents which must
happen alike to all men. But the death of a *species* is so remarkable an event
in natural history, that it deserves commemoration, and it is with no small interest
that we learn, from the archives of the University of Oxford, the exact day and
year when the remains of the last specimen of the dodo, which had rotted in
the Ashmolean museum, were cast away. The relics, we are told, were " a
Musæo subducta, annuente Vice-cancellario aliisque curatoribus, ad ea lustranda
convocatis, die Januarii, 8ᵛᵒ., A.D., 1755." Zool. Journ., No. 12, p. 559. 1828.

† Sur quelques Ossemens, &c. Ann. des Sci., tome xxi., p. 103. Sept. 1830.

Next to the direct agency of man, his indirect influence in multiplying the numbers of large herbivorous quadrupeds of domesticated races, may be regarded as one of the most obvious causes of the extermination of species. On this, and on several other grounds, the introduction of the horse, ox, and other mammalia, into America, and their rapid propagation over that continent within the last three centuries, is a fact of great importance in natural history. The extraordinary herds of wild cattle and horses which overran the plains of South America, sprung from a very few pairs first carried over by the Spaniards; and they prove that the wide geographical range of large species in great continents does not necessarily imply that they have existed there from remote periods. Humboldt observes, in his Travels *, on the authority of Azzara, that it is believed there exist, in the Pampas of Buenos Ayres, twelve million cows and three million horses, without comprising in this enumeration the cattle that have no acknowledged proprietor. In the Llanos of Caraccas, the rich hateros, or proprietors of pastoral farms, are entirely ignorant of the number of cattle they possess. The young are branded with a mark peculiar to each herd, and some of the most wealthy owners mark as many as fourteen thousand a year. In the northern plains, from the Orinoco to the lake of Maracaybo, M. Depons reckoned that one million two hundred thousand oxen, one hundred and eighty thousand horses, and ninety thousand mules, wandered at large †. In some parts of the valley of the Mississippi, especially in the country of the Osage Indians, wild horses are immensely numerous.

The establishment of black cattle in America dates from Columbus's second voyage to St. Domingo. They there multiplied rapidly; and that island presently became a kind of nursery from which these animals were successively transported to various parts of the continental coast, and from thence into the interior. Notwithstanding these numerous exportations, in twenty-seven years after the discovery of the

<hr />

* Pers. Nar., vol. iv.   † Quarterly Review, vol. xxi., p. 335.

island, herds of four thousand head, as we learn from Oviedo, were not uncommon, and there were even some that amounted to eight thousand. In 1587, the number of hides exported from St. Domingo alone, according to Acosta's report, was thirty-five thousand four hundred and forty-four ; and in the same year there were exported sixty-four thousand three hundred and fifty from the ports of New Spain. This was in the sixty-fifth year after the taking of Mexico, previous to which event the Spaniards, who came into that country, had not been able to engage in anything else than war *.

All our readers are aware that these animals are now established throughout the American continent, from Canada to Paraguay.

The ass has thriven very generally in the New World; and we learn from Ulloa, that in Quito they ran wild, and multiplied in amazing numbers, so as to become a nuisance. They grazed together in herds, and, when attacked, defended themselves with their mouths. If a horse happened to stray into the places where they fed, they all fell upon him, and did not cease biting and kicking till they left him dead †.

The first hogs were carried to America by Columbus, and established in the island of St. Domingo the year following its discovery in November, 1493. In succeeding years they were introduced into other places where the Spaniards settled ; and, in the space of half a century, they were found established in the New World, from the latitude of 25° north, to the 40th degree of south latitude. Sheep, also, and goats have multiplied enormously in the New World, as have also the cat and the rat, which last, as we before stated, has been imported unintentionally in ships. The dogs introduced by man, which have at different periods become wild in America, hunted in packs like the wolf and the jackal, destroying not only hogs, but the calves and foals of the wild cattle and horses.

* Quarterly Review, vol. xxi., p. 335.
† Ulloa's Voyage. Wood's Zoog., vol. i., p. 9.

Ulloa in his voyage, and Buffon on the authority of old writers, relate a fact which illustrates very clearly the principle before explained by us, of the check which the increase of one animal necessarily offers to that of another. The Spaniards had introduced goats into the island of Juan Fernandez, where they became so prolific as to furnish the pirates who infested those seas with provisions. In order to cut off this resource from the buccaneers, a number of dogs were turned loose into the island; and so numerous did they become in their turn, that they destroyed the goats in every accessible part, after which the number of the wild dogs again decreased *.

As an example of the rapidity with which a large tract may become peopled by the offspring of a single pair of quadrupeds, we may mention, that in the year 1773 thirteen rein-deer were exported from Norway, only three of which reached Iceland. These were turned loose into the mountains of Guldbringè Syssel, where they multiplied so greatly, in the course of forty years, that it was not uncommon to meet with herds consisting of from forty to one hundred in various districts.

In Lapland, observes a modern writer, the rein-deer is a loser by his connexion with man, but Iceland will be this creature's paradise. There is, in the interior, a tract which Sir G. Mackenzie computes at not less than forty thousand square miles, without a single human habitation, and almost entirely unknown to the natives themselves. There are no wolves ; the Icelanders will keep out the bears; and the rein-deer, being almost unmolested by man, will have no enemy whatever, unless it has brought with it its own tormenting gad-fly †.

Besides the quadrupeds before enumerated by us, our domestic fowls have also succeeded in the West Indies and America, where they have the common fowl, the goose, the duck, the peacock, the pigeon, and the guinea-fowl. As these were often taken suddenly from the temperate to very hot regions, they were not reared at first without much difficulty ; but after a

---

* Buffon, vol. v., p. 100.   Ulloa's Voyage, vol. ii., p. 220.
    † Travels in Iceland in 1810, p. 342.

few generations they became familiarized to the climate, which, in many cases, approached much nearer than that of Europe to the temperature of their original native countries.

The fact of so many millions of wild and tame individuals of our domestic species, almost all of them the largest quadrupeds and birds, having been propagated throughout the new continent within the short period that has elapsed since the discovery of America, while no appreciable improvement can have been made in the productive powers of that vast continent, affords abundant evidence of the extraordinary changes which accompany the diffusion and progressive advancement of the human race over the globe. That it should have remained for us to witness such mighty revolutions is a proof, even if there was no other evidence, that the entrance of man into the planet is, comparatively speaking, of extremely modern date, and that the effects of his agency are only beginning to be felt.

A modern writer has estimated, that there are in America upwards of four million square miles of useful soil, each capable of supporting two hundred persons ; and nearly six million, each mile capable of supporting four hundred and ninety persons *. If this conjecture be true, it will follow, as that author observes, that if the natural resources of America were fully developed, it would afford sustenance to five times as great a number of inhabitants as the entire mass of human beings existing at present upon the globe. The new continent, he thinks, though less than half the size of the old, contains an equal quantity of useful soil, and much more than an equal amount of productive power. Be this as it may, we may safely conclude that the amount of human population now existing, constitutes but a small proportion of that which the globe is capable of supporting, or which it is destined to sustain at no distant period, by the rapid progress of society, especially in America, Australia, and certain parts of the old continent.

But if we reflect that already many millions of square miles of the most fertile land, occupied originally by a boundless

* Maclaren, Art. America. Encyc. Britannica.

variety of animal and vegetable forms, have been already brought under the dominion of man, and compelled, in a great measure, to yield nourishment to him, and to a limited number of plants and animals which he has caused to increase, we must at once be convinced, that the annihilation of a multitude of species has already been effected, and will continue to go on hereafter, in certain regions, in a still more rapid ratio, as the colonies of highly-civilized nations spread themselves over unoccupied lands.

Yet, if we wield the sword of extermination as we advance, we have no reason to repine at the havoc committed, nor to fancy, with the Scotch poet, that " we violate the social union of nature ;" or complain, with the melancholy Jaques, that we

> Are mere usurpers, tyrants, and, what's worse,
> To fright the animals, and to kill them up
> In their assign'd and native dwelling-place.

We have only to reflect, that in thus obtaining possession of the earth by conquest, and defending our acquisitions by force, we exercise no exclusive prerogative. Every species which has spread itself from a small point over a wide area, must, in like manner, have marked its progress by the diminution, or the entire extirpation, of some other, and must maintain its ground by a successful struggle against the encroachments of other plants and animals. That minute parasitic plant, called " the rust " in wheat, has, like the Hessian fly, the locust, and the aphis, caused famines ere now amongst the " lords of the creation." The most insignificant and diminutive species, whether in the animal or vegetable kingdom, have each slaughtered their thousands, as they disseminated themselves over the globe, as well as the lion, when first it spread itself over the tropical regions of Africa.

We cannot conclude this division of our subject without observing, that although we have as yet considered one class only of the causes (the organic) whereby species may become exterminated, yet the continued action of these alone, throughout myriads of future ages, must work an entire change in the state of the organic creation, not merely on the continents and islands, where the power of man is chiefly exerted, but in the great ocean, where his control is almost unknown. The mind is prepared by the contemplation of such future revolutions to look for the signs of others, of an analogous nature, in the monuments of the past. Instead of being astonished at the proofs there manifested of endless mutations in the animate world, they will appear to one who has thought profoundly on the fluctuations now in progress, to afford evidence in favour of the uniformity of the system, unless, indeed, we are precluded from speaking of *uniformity* when we characterize a principle of endless variation.

CHAPTER X.

Influence of inorganic causes in changing the habitations of speecis—Powers of diffusion indispensable, that each species may maintain its ground—How changes in the physical geography affect the distribution of species—Rate of the change of species cannot be uniform, however regular the action of the inorganic causes—Illustration derived from subsidences by earthquakes—from the elevation of land by the same—from the formation of new islands—from the wearing through of an isthmus—Each change in the physical geography of large regions must occasion the extinction of species—Effects of a general alteration of climate on the migration of species—Gradual refrigeration causes species in the northern and southern hemispheres to become distinct—Elevation of temperature the reverse—Effects in the distribution of species which must result from vicissitudes in climate inconsistent with the theory of transmutation.

HAVING shown in the last chapter how considerably the numerical increase or the extension of the geographical range of any one species must derange the numbers and distribution of others, let us now direct our attention to the influence which the inorganic causes described in our first volume are continually exerting on the habitations of species.

So great is the instability of the earth's surface, that if Nature were not continually engaged in the task of sowing seeds and colonizing animals, the depopulation of a certain portion of the habitable sea and land would in a few years be considerable. Whenever a river transports sediment into a lake or sea, the aquatic animals and plants which delight in deep water are expelled: the tract, however, is not allowed to remain useless, but is soon peopled by species which require more light and heat, and thrive where the water is shallow. Every addition made to the land by the encroachment of the delta of a river banishes many subaqueous species from their native abodes; but the new-formed plain is not permitted to lie unoccupied, being instantly covered with terrestrial vegetation. The ocean devours continuous lines of sea-coast, and precipitates forests or rich pasture-land into the waves; but this space is not lost to the animate creation, for shells and sea-weed soon adhere to the new-made cliffs, and numerous fish people the channel which the current has scooped out for itself. No sooner has a volcanic isle been thrown up than some lichens begin to grow upon it, and it is sometimes clothed with verdure, while smoke and ashes are still occasionally thrown from the crater. The cocoa, pandanus, and mangrove take root upon the coral reef before it has fairly risen above the waves. The burning stream of lava that descends from Etna rolls through the stately forest, and converts to ashes every tree and herb which stand in its way; but the black strip of land thus deso-lated, is covered again, in the course of time, with oaks, pines, and chestnuts, as luxuriant as those which the fiery torrent swept away.

Every flood and landslip, every wave which a hurricane or earthquake throws upon the shore, every shower of volcanic dust and ashes which buries a country far and wide to the depth of many feet, every advance of the sand-flood, every conversion of salt-water into fresh when rivers alter their main channel of discharge, every permanent variation in the rise or fall of tides in an estuary—these and countless other causes displace in the course of a few centuries certain plants and animals from stations which they previously occupied. If, therefore, the Author of Nature had not been prodigal of those numerous contrivances before alluded to, for spreading all classes of organic beings over the earth—if he had not ordained that the fluctuations of the animate and inanimate creation should be in perfect harmony with each other, it is evident that considerable spaces, now the most habitable on the globe, would soon be as devoid of life as are the Alpine snows, or the dark abysses of the ocean, or the moving sands of the Sahara.

The powers then of migration and diffusion conferred on animals and plants, are indispensable to enable them to main-tain their ground, and would be necessary even though it were never intended that a species should gradually extend its geographical range.   But a facility of shifting their quarters being once given, it cannot fail to happen that the inhabitants of one province should occasionally penetrate into some other, since the strongest of those barriers which we before described as separating distinct regions, are all liable to be thrown down one after the other, during the vicissitudes of the earth's surface.

The numbers and distribution of particular species are affected in two ways, by changes in the physical geography of the earth.   First, these changes promote or retard the migrations of species; secondly, they alter the physical conditions of the localities which species inhabit.   If the ocean should gradually wear its way through an isthmus, like that of Suez, it would open a passage for the intermixture of the aquatic tribes of two seas previously disjoined, and would, at the same time, close a free communication which the terrestrial plants and animals of two continents had before enjoyed.   These would be, perhaps, the most important consequences in regard to the distribution of species, which would result from the breach made by the sea in such a spot ; but there would be others of a distinct nature, such as the conversion of a certain tract of land which formed the isthmus into sea.   This space previously occupied by terrestrial plants and animals would be immediately delivered over to the aquatic, a local revolution which might have happened in innumerable other parts of the globe, without being attended by any alteration in the blending together of species of two distinct provinces.

This observation leads us to point out one of the most interesting conclusions to which we are led by the contemplation of the vicissitudes of the inanimate world in relation to those of the animate.   It is clear that if the agency of inorganic causes be uniform as we have supposed, they must operate very irregularly on the state of organic beings, so that the rate according to which these will change in particular regions will not be equal in equal periods of time.

We are not about to advocate the doctrine of general catastrophes recurring at certain intervals, as in the ancient oriental cosmogonies, nor do we doubt that if very considerable periods of equal duration could be taken into our consideration and compared one with another, the rate of change in the living, as well as in the inorganic world, would be nearly uniform ; but if we regard each of the causes separately, which we know to be at present the most instrumental in remodelling the state of the surface, we shall find that we must expect each to be in action for thousands of years, without producing any extensive alterations in the habitable surface, and then to give rise, during a very brief period, to important revolutions.

We shall illustrate this principle by a few of the most remarkable examples which present themselves.   In the course of the last century, as we have before pointed out, a considerable number of instances are recorded of the solid surface, whether covered by water or not, having been permanently sunk or upraised by the power of earthquakes.   Most of these convulsions are only accompanied by temporary fluctuations in the state of limited districts, and a continued repetition of these events for thousands of years might not produce any decisive change in the state of many of those great zoological or botanical provinces of which we have sketched the boundaries.

When, for example, large parts of the ocean and even of inland seas are a thousand fathoms or upwards in depth it is a matter of no moment to the animate creation that vast tracts should be heaved up many fathoms at certain intervals, or should subside to the same amount.   Neither can any material revolution be produced in South America either in the terrestrial or the marine plants or animals by a series of shocks on the coast of Chili, each of which, like that of Penco, in 1750, should uplift the coast about twenty-five feet.   Nor if the ground sinks fifty feet at a time, as in the harbour of Port Royal, in Jamaica, in 1692, will such alterations of level work any general fluctuations in the state of organic beings inhabiting the West India islands, or the Caribbean Sea.

It is only when these subterranean powers, by shifting

gradually the points where their principal force is developed, happen to strike upon some particular region where a slight change of level immediately affects the distribution of land and water, or the state of the climate, or the barriers between distinct groups of species over extensive areas, that the rate of fluctuation becomes accelerated, and may, in the course of a few years or centuries, work mightier changes than had been experienced in myriads of antecedent years.

Thus, for example, a repetition of subsidences causing the narrow isthmus of Panamá to sink down a few hundred feet, might in a few centuries bring about a great revolution in the state of the animate creation in the western hemisphere. Thousands of aquatic species would pass for the first time from the Caribbean Sea into the Pacific; and thousands of others, before peculiar to the Pacific ocean, would make their way into the Caribbean Sea, the Gulf of Mexico, and the Atlantic. A considerable modification would probably be occasioned by the same event in the direction or volume of the Gulf-stream, and thereby the temperature of the sea and the contiguous lands would be altered as far as the influence of that current extends. A change of climate might thus be produced in the ocean from Florida to Spitzbergen, and in many countries of North America, Europe, and Greenland. Not merely the heat, but the quantity of rain which falls would be altered in certain districts, so that many species would be excluded from tracts where they before flourished; others would be reduced in number; and some would thrive more and multiply. The seeds also and the fruits of plants would no longer be drifted in precisely the same directions, nor the eggs of aquatic animals; neither would species be any longer impeded in their migrations towards particular stations before shut out from them by their inability to cross the mighty current.

Let us take another example from a part of the globe which is at present liable to suffer by earthquakes, viz., the low sandy tract which intervenes between the sea of Azof and the

Caspian. If there should occur a sinking down to a trifling amount, and such ravines should be formed as might be produced by a few earthquakes, not more considerable than have fallen within our limited observation during the last one hundred and forty years, the waters of the sea of Azof would pour rapidly into the Caspian, which, according to the lowest estimate, is fifty feet lower than the level of the Black Sea, and which, according to some writers of considerable authority, is one hundred and fifty feet,—according to others, three hundred feet below the level of the Sea of Azof *. The latter sea would immediately borrow from the Euxine, the Euxine from the Mediterranean, and the Mediterranean from the Atlantic, so that an inexhaustible current would pour down into the low tracts of Asia bordering the Caspian, by which all the sandy salt steppes adjacent to that sea would be inundated.

The diluvial waters would reach the salt lake of Aral, nor stop until their eastern shores were bounded by the high land which in the steppe of the Kirghis connects the Altay with the Himalaya mountains. A few years, perhaps a few months might suffice for the accomplishment of this great revolution in the geography of the interior of Asia; and it is impossible for those who believe in the permanence of the energy with which existing causes now act, not to anticipate such events again and again in the course of future ages.

Let us next imagine a few cases of the elevation of land of small extent at certain critical points, as, for example, in the shallowest parts of the Straits of Gibraltar, where the soundings from the African to the European side give only two hundred and twenty fathoms. In proportion as this submarine barrier of rock was upheaved, to effect which would merely require the shocks of partial and confined earthquakes, the volume of water which pours in from the Atlantic into the Mediterranean would be lessened. But the loss of the inland sea by evaporation would remain the same, so that being no

* Malte-Brun, vol. vi. p. 405.

M 2

longer able to draw on the ocean for a supply sufficient to restore its equilibrium, it must sink, and leave dry a certain portion of land around its borders. The current which now flows constantly out of the Black Sea into the Mediterranean would then rush in more rapidly, and the level of the Mediterranean would be thereby prevented from falling so low; but the level of the Black Sea would, for the same reason, sink, so that when, by a continued series of elevatory movements, the Straits of Gibraltar had become completely closed up, we might expect large and level sandy steppes to surround both the Euxine and Mediterranean, like those occurring at present on the skirts of the Caspian, and the sea of Aral. The geographical range of hundreds of aquatic species would be thereby circumscribed, and that of hundreds of terrestrial plants and animals extended.

A line of submarine volcanos crossing the channel of some strait, and gradually choking it up with ashes and lava, might produce a new barrier as effectually as a series of earthquakes; especially if thermal springs, plentifully charged with carbonate of lime, silica, and other mineral ingredients, should promote the rapid multiplication of corals and shells, and cement them together with solid matter precipitated during the intervals between eruptions. Suppose in this manner a stoppage to be caused of the Bahama Channel between the bank of that name and the coast of Florida. This insignificant revolution, confined to a mere spot in the bottom of the ocean, would, by diverting the main current of the Gulf-stream, give rise more effectually than the opening of the Straits of Panamá before supposed, to extensive changes in the climate and distribution of animals and plants inhabiting the northern hemisphere.

A repetition of elevatory movements of earthquakes might continue over an area as extensive as Europe, for thousands of ages, at the bottom of the ocean in certain regions, and produce no visible effects; whereas, if they should operate in some shallow parts of the Pacific, amid the coral archipelagos, they would soon give birth to a new continent. Hundreds of volcanic islands may be thrown up and become covered with vegetation, without causing more than local fluctuations in the animate world; but if a chain like the Aleutian archipelago or the Kurile isles, run for a distance of many hundred miles, so as to form an almost uninterrupted communication between two continents, or two distant islands, the migrations of plants, birds, insects, and even of some quadrupeds, may cause in a short time an extraordinary series of revolutions, tending to augment the range of some animals and plants, and to limit that of others. A new archipelago might be formed in the Mediterranean, the Bay of Biscay, and a thousand other localities, and might produce less important events than one rock which should rise up between Australia and Java, so placed that winds and currents might cause an interchange of the plants, insects, and birds, of the latter countries.

If we turn from the igneous to the aqueous agents, we find the same tendency to an irregular rate of change, naturally connected with the strictest uniformity in the energy of those causes. When the sea, for example, gradually encroaches upon both sides of a narrow isthmus, as that of Sleswick, separating the North Sea from the Baltic, where, as we stated, the cliffs on both the opposite coasts are wasting away *, no material alteration results for thousands of years, save only that there is a progressive conversion of a small strip of land into water. A few feet only, or a few yards, are annually removed; but when at last the partition shall be broken down, and the tides of the ocean shall enter by a direct passage into the inland sea, instead of going by a circuitous route through the Cattegat, a body of salt-water will sweep up as far as the Gulfs of Bothnia and Finland, the waters of which are now brackish, or almost fresh; and this revolution will be attended by the local annihilation of many species.

Similar consequences must have resulted, on a small scale, when the sea opened its way through the isthmus of Staveren

* Vol. i. p. 289.

in the thirteenth century, forming an union between an inland
lake and the ocean, and opening, in the course of one century,
a shallow strait more than half as wide as the narrowest part
of that which divides England from France.

It will almost seem superfluous, after we have thus traced
the important modifications in the condition of living beings
which flow from changes of trifling extent, to argue that entire
revolutions might be brought about, if the climate and physical
geography of the whole globe were greatly altered. Species we
have stated are, in general, local, some being confined to extremely small spots, and depending for their existence on a combination of causes which, if they are to be met with elsewhere,
occur only in some very remote region. Hence it must happen
that when the nature of these localities is changed the species
will perish ; for it will rarely happen that the cause which
alters the character of the district will afford new facilities to
the species to establish itself elsewhere.

If we attribute the origin of a great part of the desert of
Africa to the gradual progress of moving sands, driven eastward by the westerly winds, we may safely infer that a variety
of species must have been annihilated by this cause alone. The
sand-flood has been inundating, from time immemorial, the
rich lands on the west of the Nile, and we have only to multiply this effect a sufficient number of times, in order to understand how, in the lapse of ages, a whole group of terrestrial
animals and plants may become extinct.

This desert, without including Bornou and Darfour, extends,
according to the calculation of Humboldt, over one hundred
and ninety-four thousand square leagues, an area far more than
double that of the Mediterranean, which occupies only seventy-
nine thousand eight hundred square leagues. In a small portion of so vast a space, we may infer, from analogy, that there
were many peculiar species of plants and animals which must
have been banished by the sand, and their habitations invaded
by the camel and by birds and insects formed for the arid
sands.

There is evidently nothing in the nature of the catastrophe to favour the escape of the former inhabitants to some adjoining province; nothing to weaken, in the bordering lands, that powerful barrier against emigration—pre-occupancy. Nor, even if the exclusion of a certain group of species from a given tract were compensated by an extension of their range over a new country, would that circumstance tend to the conservation of species in general; for the extirpation would merely then be transferred to the region so invaded. If it be imagined, for example, that the aboriginal quadrupeds, birds, and other animals of Africa emigrated in consequence of the advance of drift-sand, and colonized Arabia, the indigenous Arabian species must have given way before them, and have been reduced in number or destroyed.

Let us next suppose that, in some central and more elevated parts of the great African desert, the upheaving power of earthquakes should be exerted throughout an immense series of ages, accompanied, at certain intervals, by volcanic eruptions such as gave rise at once, in 1755, to a mountain one thousand seven hundred feet high, on the Mexican plateau. When the continued repetition of these events had caused a mountain-chain, it is obvious that a complete transformation in the state of the climate would be brought about throughout a vast area.

We will imagine the summits of the new chain to rise so as to be covered, like Mount Atlas, for several thousand feet, with snow during a great part of the year. The melting of these snows, during the greatest heat, would cause the rivers to swell in the season when the greatest drought now prevails; the waters, moreover, derived from this source, would always be of lower temperature than the surrounding atmosphere, and would thus contribute to cool the climate. During the numerous earthquakes and volcanic eruptions which would attend the gradual formation of the chain, there would be many floods, caused by the bursting of temporary lakes and by the melting of snows by lava. These inundations would deposit alluvial matter far and wide over the original sands at all levels, as the country assumed various shapes, and was modified again and again by the moving power from below, and the aqueous erosion of the surface above. At length the Sahara would be fertilized, irrigated by rivers and streamlets intersecting it in every direction, and covered by jungle and morasses, so that the animals and plants which now people northern Africa would disappear, and the region would gradually become fitted for the reception of a population of species perfectly dissimilar in their forms, habits, and organization.

There are always some peculiar and characteristic features in the physical geography of each large division of the globe ; and on these peculiarities the state of animal and vegetable life is dependent. If, therefore, we admit incessant fluctuations in the physical geography, we must, at the same time, concede the successive extinction of terrestrial and aquatic species to be part of the economy of our system. When some great class of stations is in excess in certain latitudes, as, for example, in wide savannahs, arid sands, lofty mountains, or inland seas, we find a corresponding development of species adapted for such circumstances. In North America, where there is a chain of vast inland lakes of fresh-water, we find an extraordinary abundance and variety of aquatic birds, fresh-water fish, testacea, and small amphibious reptiles, fitted for such a climate. The greater part of these would perish if the lakes were destroyed,—an event that might be brought about by some of the least of those important revolutions contemplated in geology. It might happen that no fresh-water lakes of corresponding magnitude might then exist on the globe ; but if they occurred elsewhere, they might be situated in New Holland, Southern Africa, Eastern Asia, or some region so distant as to be quite inaccessible to the North American species ; or they might be situated within the tropics, in a climate uninhabitable by species fitted for a temperate zone ; or, finally, we may presume that they would be pre-occupied by *indigenous* tribes.

To pursue this train of reasoning farther is unnecessary ; the reader has only to reflect on what we have said of the habitations and the stations of organic beings in general, and to consider them in relation to those effects which we have contemplated in our first volume as resulting from the igneous and aqueous causes now in action, and he will immediately perceive that, amidst the vicissitudes of the earth's surface, species cannot be immortal, but must perish one after the other, like the individuals which compose them. There is no possibility of escaping from this conclusion, without resorting to some hypothesis as violent as that of Lamarck, who imagined, as we have before seen, that species are each of them endowed with indefinite powers of modifying their organization, in conformity to the endless changes of circumstances to which they are exposed.

Some of the effects which must attend every general alteration of *climate* are sufficiently peculiar to claim a separate consideration before concluding the present chapter.

We have before stated that, during seasons of extraordinary severity, many northern birds, and, in some countries, many quadrupeds, migrate southwards. If these cold seasons were to become frequent, in consequence of a gradual and general refrigeration of the atmosphere, such migrations would be more and more regular, until, at length, many animals, now confined to the arctic regions, would become the tenants of the temperate zone ; while the inhabitants of the latter would approach nearer to the equator. At the same time, many species previously established on high mountains, would begin to descend, in every latitude, towards the middle regions, and those which were confined to the flanks of mountains would make their way into the plains. Analogous changes would also take place in the vegetable kingdom.

If, on the contrary, the heat of the atmosphere be on the increase, the plants and animals of low grounds would ascend to higher levels, the equatorial species would migrate into the temperate zone, and those of the temperate into the arctic circle.

But although some species might thus be preserved, every

great change of climate must be fatal to many which can find
no place of retreat, when their original habitations become
unfit for them. For if the general temperature be on the
rise, then is there no cooler region whither the polar species
can take refuge; if it be on the decline, then the animals and
plants previously established between the tropics have no
resource. Suppose the general heat of the atmosphere to increase, so that even the arctic region became too warm for the
musk-ox and rein-deer, it is clear that they must perish; so,
if the torrid zone should lose so much of its heat by the progressive refrigeration of the earth's surface, as to be an unfit
habitation for apes, boas, bamboos, and palms, these tribes of
animals and plants, or at least most of the species now belonging to them, would become extinct, for there would be no
warmer latitudes for their reception.

It will follow, therefore, that as often as the climates of the
globe are passing from the extreme of heat to that of cold—
from the summer to the winter of the great year before alluded
to by us *—the migratory movement will be directed constantly
from the poles towards the equator; and for this reason the
species inhabiting parallel latitudes, in the northern and
southern hemispheres, must become widely different. For
we assume, on grounds before stated †, that the original stock
of each species is introduced into one spot of the earth only,
and, consequently, no species can be at once indigenous in the
arctic and antarctic circles.

But when, on the contrary, a series of changes in the physical geography of the globe, or any other supposed cause,
occasions an elevation of the general temperature,—when there
is a passage from the winter to one of the vernal or summer
seasons of the great cycle of climates, then the order of the
migratory movement is inverted. The different species of
animals and plants direct their course from the equator towards
the poles; and the northern and southern hemispheres may
become peopled, to a great degree, by identical species. Such

is not the actual state of the inhabited earth, as we have already shown in our sketch of the geographical distribution of its living productions ; and this fact adds one more additional proof to a great body of evidence, derived from independent sources, that the general temperature has been cooling down during the epochs which immediately preceded our own.

We do not mean to speculate on the entire transposition of a group of animals and plants from tropical to polar latitudes, or the reverse, as a probable, or even possible, event; for although we believe the mean annual temperature of one zone to be transferrible to another, we know that the same climate cannot be so transferred.   Whatever be the general temperature of the earth's surface, comparative equability of heat will characterize the tropical regions, while great periodical variations will belong to the temperate, and still more to the polar latitudes.   These, and many other peculiarities connected with heat and light, depend on fixed astronomical causes, such as the motion of the earth and its position in relation to the sun, and not on those fluctuations of its surface, which may influence the general temperature.

Among many obstacles to such extensive transferences of habitations, we must not forget the immense lapse of time required, according to any hypothesis yet suggested, especially that which has appeared to us most feasible, to bring about a considerable change in climate.   During a period so vast, the other causes of extirpation, before enumerated by us, would exert so powerful an influence as to prevent all, save a very few hardy species, from passing from equatorial to polar regions, or from the tropics to the pole.

But the power of accommodation to new circumstances is great in certain species, and might enable many to pass from one zone to another, if the mean annual heat of the atmosphere and the ocean were greatly altered.   To the marine tribes, especially, such a passage would be possible, for they are less impeded in their migrations, by barriers of land, than are the terrestrial by the ocean.   Add to this, that the temperature of the ocean is much more uniform than that of the atmosphere investing the land, so that we may easily suppose that most of the testacea, fish, and other classes, might pass from the equatorial into the temperate regions, if the mean temperature of those regions were transposed, although a second expatriation of these species of tropical origin into the arctic and antarctic circles would probably be impossible.

On the principles above explained, if we found that at some former period, as when, for example, our carboniferous strata were deposited, the same tree-ferns and other plants inhabited the regions now occupied by Europe and Van Dieman's Land, we should suspect that the species in question had, at some antecedent period, inhabited lands within the tropics, and that an increase of the mean annual heat had caused them to emigrate into both the temperate zones.   There are no geological data, however, as yet obtained, to warrant the opinion that such identity of species existed in the two hemispheres in the era in question.

Let us now consider more particularly the effect of vicissitudes of climate in causing one species to give way before the increasing numbers of some other.

When temperature forms the barrier which arrests the progress of an animal or plant in a particular direction, the individuals are fewer and less vigorous as they approach the extreme confines of the geographical range of the species. But these stragglers are ready to multiply rapidly on the slightest increase or diminution of heat that may be favourable to them, just as particular insects increase during a hot summer, and certain plants and animals gain ground after a series of congenial seasons.

In almost every district, especially if it be mountainous, there are a variety of species the limits of whose habitations are conterminous, some being unable to proceed farther without encountering too much heat, others too much cold.   Individuals, which are thus on the borders of the regions proper to their respective species, are like the out-posts of hostile armies, ready to profit by every slight change of circumstances in their favour, and to advance upon the ground occupied by their neighbours and opponents.

The proximity of distinct climates, produced by the inequalities of the earth's surface, brings species possessing very different constitutions into such immediate contact, that their naturalizations are very speedy whenever opportunities of advancing present themselves.   Many insects and plants, for example, are common to low plains within the arctic circle, and to lofty mountains in Scotland and other parts of Europe. If the climate, therefore, of the polar regions were transferred to our own latitudes, the species in question would immediately descend from these elevated stations to overrun the low grounds.   Invasions of this kind, attended by the expulsion of the pre-occupants, are almost instantaneous, because the change of temperature not only places the one species in a more favourable position, but renders the others sickly and almost incapable of defence.

Lamarck appears to have speculated on the modifications to which every variation of external circumstances might give rise in the form and organization of species, as if he had indefinite periods of time at his command, not sufficiently reflecting that revolutions in the state of the habitable earth, whether by changes of climate or any other condition, are attended by still greater fluctuations in the relative condition of contemporary species.   They can avail themselves of these alterations in their favour instantly, and augment their numbers to the injury of some other species; whereas the supposed transmutations are only assumed to be brought about by slow and insensible degrees, and in a lapse of ages, the duration of which is beyond the reach of human conception   Even if we thought it possible that the palm or the elephant, which now flourish in equatorial regions, could ever learn to bear the variable seasons of our temperate zone, or the rigours of an arctic winter, we should, with no less confidence, affirm, that they must perish before they had time to become habituated to such new circumstances.

That they would be supplanted by other species at each variation of climate, may be inferred from what we have before said
of the known local exterminations of species which have resulted
from the multiplication of others. Some minute insect, perhaps, might be the cause of destruction to the huge and
powerful elephant.

Suppose the climate of the highest part of the woody zone
of Etna to be transferred to the sea-shore at the base of the
mountain, no botanist would anticipate that the olive, lemon-
tree, and prickly pear (*Cactus opuntia*), would be able to contend with the oak and chestnut, which would begin forthwith
to descend to a lower level, or that these last would be able to
stand their ground against the pine, which would also, in the
space of a few years, begin to occupy a lower position. We
might form some kind of estimate of the time which might be
required for the migrations of these plants; whereas we have
no data for concluding that any number of thousands of years
would be sufficient for one step in the pretended metamorphosis of one species into another, possessing distinct attributes
and qualities.

This argument is applicable not merely to *climate*, but to
any other cause of mutation. However slowly a lake may be
converted into a marsh, or a marsh into a meadow, it is evident
that before the lacustrine plants can acquire the power of living
in marshes, or the marsh-plants of living in a less humid soil,
other species, already existing in the region, and fitted for
these several stations, will intrude and keep possession of the
ground. So if a tract of salt-water becomes fresh by passing
through every intermediate degree of brackishness, still the
marine molluscs will never be permitted to be gradually metamorphosed into fluviatile species; because long before any such
transformation can take place by slow and insensible degrees,
other tribes, which delight in brackish or fresh-water, will avail
themselves of the change in the fluid, and will, each in their
turn, monopolize the space.

It is idle to dispute about the abstract possibility of the conversion of one species into another, when there are known causes so much more active in their nature, which must always intervene and prevent the actual accomplishment of such conversions. A faint image of the certain doom of a species less fitted to struggle with some new condition in a region which it previously inhabited, and where it has to contend with a more vigorous species, is presented by the extirpation of savage tribes of men by the advancing colony of some civilized nation. In this case the contest is merely between two different *races*, each gifted with equal capacities of improvement—between two varieties, moreover, of a species which exceeds all others in its aptitude to accommodate its habits to the most extraordinary variations of circumstances. Yet few future events are more certain than the speedy extermination of the Indians of North America and the savages of New Holland in the course of a few centuries, when these tribes will be remembered only in poetry and tradition.

CHAPTER XI.

Theory of the successive extinction of species consistent with their limited geographical distribution—The discordance in the opinions of botanists respecting
the centres from which plants have been diffused may arise from changes in
physical geography subsequent to the origin of living species—Whether there
are grounds for inferring that the loss from time to time of certain animals and
plants is compensated by the introduction of new species?—Whether any
evidence of such new creations could be expected within the historical era, even
if they had been as frequent as cases of extinction?—The question whether
the existing species have been created in succession can only be decided by
reference to geological monuments.

WE have pointed out in the preceding chapters the strict
dependence of each species of animal and plant on certain
physical conditions in the state of the earth's surface, and on
the number and attributes of other organic beings inhabiting
the same region. We have also endeavoured to show that all
these conditions are in a state of continual fluctuation, the
igneous and aqueous agents remodelling, from time to time,
the physical geography of the globe, and the migrations of
species causing new relations to spring up successively between
different organic beings. We have deduced as a corollary,
that the species existing at any particular period must, in the
course of ages, become extinct one after the other. " They
must die out," to borrow an emphatical expression from Buffon,
" because Time fights against them."

If the views which we have taken are just, there will be no
difficulty in explaining why the habitations of so many species
are now restrained within exceedingly narrow limits. Every
local revolution, such as those contemplated in the preceding
chapter, tends to circumscribe the range of some species, while
it enlarges that of others; and as we have been led to infer
that new species originate in one spot only, each must require
time to diffuse itself over a wide area. The recent origin,

therefore, of some species, and the high antiquity of others, may be equally consistent with the general fact of their limited distribution, some being local, because they have not existed long enough to admit of their wide dissemination ; others, because circumstances in the animate or inanimate world have occurred to restrict the range which they may once have obtained.

As considerable modifications in the relative levels of land and sea have taken place, in certain regions, since the existing species were in being, we can feel no surprise that the zoologist and botanist have hitherto found it difficult to refer the geographical distribution of species to any clear and determinate principles, since they have usually speculated on the phenomena, upon the assumption that the physical geography of the globe had undergone no material alteration since the introduction of the species now living.   So long as this assumption was made, the facts relating to the geography of plants and animals appeared capricious in the extreme, and by many the subject was pronounced to be so full of mystery and anomalies, that the establishment of a satisfactory theory was hopeless.

Some botanists conceived, in accordance with the hypothesis of Willdenow, that mountains were the centres of creation from which the plants now inhabiting large continents have radiated, to which Decandolle and others, with much reason, objected, that mountains, on the contrary, are often the barriers between two provinces of distinct vegetation.   The geologist who is acquainted with the extensive modifications which the surface of the earth has undergone in very recent geological epochs, may be able, perhaps, to reconcile both these theories in their application to different regions.

A lofty range of mountains, which is so ancient as to date from a period when the species of animals and plants differed from those now living, will naturally form a barrier between contiguous provinces ; but a chain which has been raised, in great part, within the epoch of existing species, and around which new lands have arisen from the sea within that period, will be a centre of peculiar vegetation.

" In France," observes Decandolle*, " the Alps and Ce-
vennes prevent a great number of the plants of the south from
spreading themselves to the northward ; but it has been remarked that some species have made their way through the
gorges of these chains, and are found on their northern sides,
principally in those places where they are lower and more
interrupted." Now the chains here alluded to have probably
been of considerable height, even since the era when the existing vegetation began to appear, and were it not for the deep
fissures which divide them, they might have caused much more
abrupt terminations to the extension of distinct assemblages of
species.

Parts of the Italian peninsula, on the other hand, have
gained a considerable portion of their present height since a
majority of the marine species now inhabiting the Mediterranean, and probably, also, since the terrestrial plants of the same
region, were in being. Large tracts of land have been added,
both on the Adriatic and Mediterranean side, to what originally constituted a much narrower range of mountains, if not
a chain of islands running nearly north and south, like Cor-
sica and Sardinia. It may, therefore, be presumed, that the
Apennines have been a centre whence species have diffused
themselves over the contiguous *lower* and *newer* regions. In
this and all analogous situations, the doctrine of Willdenow,
that species have radiated from the mountains as from centres,
may be well founded.

It appears from Mr. Brown's remarks on the vegetation of
New Holland, that there are two groups of plants occurring
between the thirty-third and thirty-fifth degrees of southern
latitude, and principally at the two opposite extremities of this
tract, that is, near the eastern and western coasts. These points
have been termed the two principal *foci* of Australian vegetation, each of them possessing certain genera which are almost
peculiar to it †. If, when this continent has been more tho-

---

* Essai Elémentaire, &c., p. 46.
† Brown's Appendix to Flinders's Voyage, and Prichard, Phys. Hist. of Mankind, vol. i., p. 31.

roughly investigated, we do not discover some physical barriers, such as a great marsh, or a desert, or a lofty mountain-chain, now intervening between these districts, there may, perhaps, be geological evidence hereafter discovered, that a sea was interposed up to a modern period separating two large islands. Sufficient time may not have elapsed since the union of such isles, to allow of a complete intermixture by mutual immigrations.

If the reader should infer, from the facts laid before him in the preceding chapters, that the successive extinction of animals and plants may be part of the constant and regular course of nature, he will naturally inquire whether there are any means provided for the repair of these losses? Is it part of the economy of our system that the habitable globe should, to a certain extent, become depopulated both in the ocean and on the land ; or that the variety of species should diminish until some new era arrives when a new and extraordinary effort of creative energy is displayed? Or is it possible that new species can be called into being from time to time, and yet that so astonishing a phenomenon can escape the observation of naturalists?

Humboldt has characterized these subjects as among the mysteries which natural science cannot reach ; and he observes, that the investigation of the origin of beings does not belong to zoological or botanical geography. To geology, however, these topics do strictly appertain ; and this science is only interested in inquiries into the state of the animate creation as it now exists, with a view of pointing out its relations to antecedent periods when its condition was different.

Before offering any hypothesis towards the solution of so difficult a problem, let us consider what kind of evidence we ought to expect, in the present state of science, of the first appearance of new animals or plants, if we could imagine the successive creation of species to constitute, like their gradual extinction, a regular part of the economy of nature.

In the first place it is obviously more easy to prove that a

N 2

species, once numerously represented in a given district, has ceased to be, than that some other which did not pre-exist has made its appearance—assuming always, for reasons before stated, that single stocks only of each animal and plant are originally created, and that individuals of new species do not suddenly start up in many different places at once.

So imperfect has the science of Natural History remained down to our own times, that within the memory of persons now living, the numbers of known animals and plants have been doubled, or even quadrupled, in many classes. New and often conspicuous species are annually discovered in parts of the old continent, long inhabited by the most civilized nations. Conscious, therefore, of the limited extent of our information, we always infer, when such discoveries are made, that the beings in question had previously eluded our research ; or had at least existed elsewhere, and only migrated at a recent period into the territories where we now find them. It is difficult even in contemplation to anticipate the time when we shall be entitled to make any other hypothesis in regard to all the marine tribes, and to by far the greater number of the terrestrial ;—such as birds, which possess such unlimited powers of migration ; insects which, besides their numbers, are also so capable of being diffused to vast distances; and cryptogamous plants, to which, as to many other classes, both of the animal and vegetable kingdom, similar observations are applicable.

What kind of proofs, therefore, could we reasonably expect to find of the origin at a particular period of a new species ?

Perhaps it may be said in reply, that within the last two or three centuries some forest tree or new quadruped might have been observed to appear suddenly in those parts of England or France which had been most thoroughly investigated ; —that naturalists might have been able to shew that no such being inhabited any other region of the globe, and that there was no tradition of anything similar having before been observed in the district where it had made its appearance.

Now although this objection may seem plausible, yet its

force will be found to depend entirely on the rate of fluctuation which we suppose to prevail in the animate world, and on the proportion which such conspicuous subjects of the animal and vegetable kingdoms bear to those which are less known, and escape our observation. There are probably more than a million, perhaps two millions of species of plants and animals, exclusive of the microscopic and infusory animalcules, now inhabiting the terraqueous globe. The terrestrial plants, it is supposed, may amount, if fully known, to about one hundred thousand, and the insects to four times that number. To these we have still to add for the remainder of the terrestrial classes, many of the invertebrated and all the vertebrated animals. As to the aquatic tribes, it remains at present in a great degree mere matter of conjecture what proportion they bear to the denizens of the land ; but the habitable surface beneath the waters can hardly be estimated at less than double that of the continents and islands, even admitting that a very considerable area is destitute of life, in consequence of great depth, cold, darkness, and other circumstances. In the late polar expedition it was found that in some regions, as in Baffin's Bay, there were marine animals inhabiting the bottom at great depths, where the temperature of the water was below the freezing point. That there is life at much greater profundities in warmer regions may be confidently inferred. We have before stated that marine plants not only exist but acquire vivid colours at depths where, to our senses, there would be darkness deep as night.

The ocean teems with life—the class of *polyps* alone are conjectured by Lamarck to be as strong in individuals as insects. Every tropical reef is described as bristling with corals, budding with sponges, and swarming with crustacea, echini, and testacea ; while almost every tide-washed rock is carpeted with fuci and studded with corallines, actiniæ, and mollusca. There are innumerable forms in the seas of the warmer zones, which have scarcely begun to attract the attention of the naturalist ; and there are parasitic animals without number, three or four of which are sometimes appropriated to one genus, as to the *Balæna,* for example. Even though we concede, therefore, that the geographical range of marine species is more extensive in general than that of the terrestrial, (the temperature of the sea being more uniform, and the land impeding less the migrations of the oceanic than the ocean those of the terrestrial,) yet we think it most probable that the aquatic species far exceed in number the inhabitants of the land.

Without insisting on this point, we may safely assume, as we before stated, that, exclusive of microscopic beings, there are between one and two millions of species now inhabiting the terraqueous globe; so that if only one of these were to become extinct annually, and one new one were to be every year called into being, more than a million of years would be required to bring about a complete revolution in organic life.

We are not hazarding at present any hypothesis as to the probable rate of change, but none will deny that when we propose as a mere speculation the *annual* birth and the *annual* death of one species on the globe, we imagine no slight degree of instability in the animate creation. If we divide the surface of the earth into twenty regions of equal area, one of these might comprehend a space of land and water about equal in dimensions to Europe, and might contain a twentieth part of the million of species which we will suppose to exist. In this region one species only would, according to the rate of mortality before assumed, perish in twenty years, or only five out of fifty thousand in the course of a century. But as a considerable proportion of the whole would belong to the aquatic classes, with which we have a very imperfect acquaintance, we may exclude them from our consideration, and thus one only might be lost in about forty years among the terrestrial tribes. Now the mammiferous quadrupeds in Great Britain are only to other terrestrial species of organic beings, both plants and animals, in the proportion of about one to two hundred and eighty; and taking this as a rude approximation to a general standard, it would require more than eight thousand years before it would come to the turn of this conspicuous class to lose one of their number even in a region of the dimensions of Europe.

It is easy, therefore, to conceive, that in a small portion of such an area, in countries, for example, of the size of England and France, periods of much greater duration must elapse before it would be possible to authenticate the first appearance of one of the larger plants and animals, assuming the annual birth and death of one species to be the rate of vicissitude in the animate creation throughout the world.

The observations of naturalists may, in the course of future centuries, accumulate positive data, from which an insight into the laws which govern this part of our terrestrial system may be derived ; but, in the present deficiency of historical records, we have traced up the subject to that point where geological monuments alone are capable of leading us on to the discovery of ulterior truths.   To these, therefore, we must now appeal, carefully examining the strata of recent formation wherein the remains of *living* species, both animal and vegetable, are known to occur.   We must study these strata in strict reference to their chronological order as deduced from their superposition, and other relations.   From these sources we may learn which of the species, now our contemporaries, have survived the greatest revolutions of the earth's surface ; which of them have co-existed with the greatest number of animals and plants now extinct, and which have made their appearance only when the animate world had nearly attained its present condition.

From such data we may be enabled to infer whether species have been called into existence in succession or all at one period ; whether singly, or whether by groups simultaneously ; whether the antiquity of man be as high as that of any of the inferior beings which now share the planet with him, or whether the human species is one of the most recent of the whole.

To some of these questions we can even now return a satisfactory answer; and with regard to the rest, we have some data to guide conjecture, and to enable us to speculate with advantage: but it would be premature to anticipate such discussions until we have laid before the reader an ample body of materials amassed by the industry of modern geologists.

———

CHAPTER XII.

THE second branch of our inquiry, respecting the changes of the organic world, relates to the effects produced by the powers of vitality on the state of the earth's surface, and on the material constituents of its crust.

By the effects produced on the surface, we mean those modifications in physical geography of which the existence of organic beings is the direct cause,—as when the growth of certain plants covers the slope of a mountain with peat, or converts a swamp into dry land; or when vegetation prevents the soil, in certain localities, from being washed away by running water.

By the agency of the powers of vitality on the material constituents of the earth's crust, we mean those permanent modifications in the composition and structure of new strata, which result from the imbedding therein of animal and vegetable remains. In this case, organic beings may not give rise immediately to any new features in the physical geography of certain tracts, which would not equally have resulted from the mere operation of inorganic causes; as, for example, if a lake be filling up with sediment, held in suspension by the waters of some river, and with mineral matter precipitated from the waters of springs, the character of the deposits may be modified by aquatic animals and plants, which may convert the earthy particles into shell, peat, and other substances : but the lake may, nevertheless, be filled up in the same time, and the new strata may be deposited in nearly the same order as would have prevailed if its waters had never been peopled by living beings.

In treating of the first division of our subject we may remark, that when we talk of alterations in physical geography, we are apt to think too exclusively of that part of the earth's surface which has emerged from beneath the waters, and with which alone, as terrestrial beings, we are familiar. Here the direct power of animals and plants to cause any important variations is, of necessity, very limited, except in checking the progress of that decay of which the land is the chief theatre. But if we extend our views, and instead of contemplating the dry land we consider that larger portion which is assigned to the aquatic tribes, we discover the immediate influence of the living creation, in imparting varieties of conformation to the solid exterior which the sole agency of inanimate causes would not produce, to be very great.

Thus, when timber is floated into the sea, it is often drifted to vast distances and subsides in spots where there might have been no deposit, at that time and place, if the earth had not been tenanted by living beings. If, therefore, in the course of ages, a hill of wood, or lignite, be thus formed in the subaqueous regions, a change in the submarine geography may be said to have resulted from the action of organic powers. So in regard to the growth of coral reefs : it is probable that almost all the matter of which they are composed is supplied by mineral springs, which we know often rise up at the bottom of the sea, and which, on land, abound throughout volcanic regions thousands of miles in extent. The matter thus constantly given out could not go on accumulating for ever in the waters, but would be precipitated in the abysses of the sea, even if there were no polyps and testacea; but these animals arrest and secrete the carbonate of lime on the summits of submarine mountains, and form reefs many hundred feet in thickness, and hundreds of leagues in length, where, but for them, none might ever have existed.

If no such voluminous masses are formed on the land, it is not from the want of solid matter in the structure of terrestrial animals and plants, but merely because, as we have so often stated, the continents are those parts of the globe where accessions of matter can scarcely ever take place,—where, on the contrary, the most solid parts already formed are, each in their turn, exposed to gradual degradation. The quantity of timber and vegetable matter which grows in a tropical forest in the course of a century is enormous, and multitudes of animal skeletons are scattered there in the same period, besides innumerable land-shells and other organic substances. The aggregate of these materials might constitute, perhaps, a mass greater in volume than that which is produced in any coral-reef during the same lapse of years; but, although this process should continue on the land for ever, no mountains of wood or bone would be seen stretching far and wide over the country, or pushing out bold promontories into the sea.

The whole solid mass is either devoured by animals, or decomposes, as does a portion of the rock and soil on which the animals and plants are supported. For the decomposition of the strata themselves, especially of their alkaline ingredients and of the organic remains which they so frequently include, is one source from whence running water and the atmosphere may derive the materials which are absorbed by the roots and leaves of plants. Another source is the passage into a gaseous form of even the hardest parts of animals and plants which die and are exposed to putrefy in the air, where they are soon resolved into the elements of which they are composed; and while a portion of these parts is volatilized, the rest is taken up by rain-water and sinks into the earth or flows towards the sea, so that they enter again and again into the composition of different organic beings.

The principal elements found in plants are hydrogen, carbon, and oxygen, so that water and the atmosphere contain all of them, either in their own composition or in solution *. The constant supply of these elements is maintained not only by the putrefaction of animal and vegetable substances, and the decay of rocks before mentioned, but also by the copious evolution of carbonic acid and other gases from volcanos and mineral springs, and by the effects of ordinary evaporation, whereby aqueous vapours are made to rise from the ocean and to circulate round the globe.

It is well known that when two gases of different specific gravity are brought into contact, even though the heavier be the lowermost, they become uniformly diffused by mutual absorption through the whole space which they occupy. By virtue of this law, the heavy carbonic acid finds its way upwards through the lighter air, and conveys nourishment to the lichen which covers the mountain top.

The fact, therefore, that the vegetable mould which covers the earth's surface does not decrease in thickness, will not altogether bear out the argument which was founded upon it by Playfair. This vegetable soil, he observes, consists partly of loose earthy materials easily removed, in the form of sand and gravel, partly of finer particles suspended in the waters, which tinge those of some rivers continually, and those of all occasionally, when they are flooded. The soil, although continually diminished from this cause, " remains the same in quantity, or at least nearly the same, and must have done so ever since the earth was the receptacle of animal or vegetable life. The soil, therefore, is augmented from other causes, just as much, at an average, as it is diminished by that now mentioned; and this augmentation evidently can proceed from nothing but the constant and slow disintegration of the rocks †."

* See some good remarks on the Formation of Soils, Bakewell's Geology, chap. xviii.

† Illust. of Hutt. Theory, § 103.

That the repair of the *earthy* portion of the soil can only proceed, as Playfair suggests, from the decomposition of rocks, may be admitted ; but the *vegetable* matter may be supplied, and is actually furnished in a great degree, by absorption from the atmosphere, as we before mentioned, so that in level situations, such as in platforms that intervene between valleys where the action of running water is very trifling, the fine vegetable particles carried off by the rain may be perpetually restored, not by the waste of the rock below, but from the air above.

If we supposed the quantity of food consumed by terrestrial animals, together with the matter absorbed by them in breathing, and the elements imbibed by the roots and leaves of plants, to be derived entirely from that supply of hydrogen, carbon, oxygen, azote, and other elements, given out into the atmosphere and the waters by the putrescence of organic substances, then we might imagine that the vegetable mould would, after a series of years, neither gain nor lose a single particle by the action of organic beings.   This conclusion is not far from the truth ; but the operation which renovates the vegetable and animal mould is by no means so simple as that here supposed. Thousands of carcasses of terrestrial animals are floated down every century into the sea, and, together with forests of drift-timber, are imbedded in subaqueous deposits, where their elements are imprisoned in solid strata, and may there remain throughout whole geological epochs before they again become subservient to the purposes of life.

On the other hand, fresh supplies are derived by the atmosphere, and by running water, as we before stated, from the disintegration of rocks and their organic contents, and from the interior of the earth, from whence all the elements before-mentioned, which enter principally into the composition of animals and vegetables, are continually evolved. Even nitrogen has been recently found to be contained very generally in the waters of mineral springs *.

* Dr. Daubeny has ascertained this interesting fact in his late tour on the continent.

If we suppose that the copious discharge from the nether regions, by springs and volcanic vents, of carbonic acid and other gases, together with the decomposition of rocks, may be just sufficient to counterbalance that loss of matter which, having already served for the nourishment of animals and plants, is annually carried down in organized forms, and buried in subaqueous strata, we believe that we concede the utmost that is consistent with probability. When more is required by a theorist—when we are told that a counterpoise is derived from the same source to that enormous disintegration of solid rock and its transportation to lower levels, which is the annual result of the action of rivers and marine currents, we must entirely withhold our assent. Such an opinion has been recently advanced by an eminent geologist, or we should have deemed it unnecessary to dwell on propositions which appear to us so clear and obvious.

The descriptions which we gave of the degradation yearly going on through the eastern shores of England, and of the enormous weight of solid matter hourly rolled down by the Ganges or the Mississippi, have been represented as extreme cases, calculated to give a partial view of the changes now in progress, especially as we omitted, it is said, to point out the silent but universal action of a great antagonist power, whereby the destructive operations before alluded to are neutralized, and even, in a great degree, counterbalanced.

" Are there," says Professor Sedgwick, " no *antagonist* powers in nature to oppose these mighty ravages—no conservative principle to meet this vast destructive agency ? The forces of degradation very often of themselves produce their own limitation. The mountain-torrent may tear up the solid rock and bear its fragments to the plain below; but there its power is at an end, and the rolled fragments are left behind to a new action of material elements. And what is true of a single rock, is true of a mountain-chain ; and vast regions on the surface of the earth, now only the monuments of spoliation and waste, may hereafter rest secure under the defence of a thick vegetable covering, and become a new scene of life and animation.

" It well deserves remark that the destructive powers of nature act only upon lines, while some of the grand principles of conservation act upon the whole surface of the land. By the processes of vegetable life an incalculable mass of solid matter is absorbed, year after year, from the elastic and non-elastic fluids circulating round the earth, and is then thrown down upon its surface. In this *single* operation there is a *vast counterpoise to all* the agents of destruction. And the deltas of the Ganges and the Mississippi are not solely formed at the expense of the solid materials of our globe, but in part, and I believe also in a considerable part, by one of the great conservative operations by which the elements are made to return into themselves *."

This is splendid eloquence, full of the energy and spirit that breathes through the whole address :—

> Monte decurrens velut amnis, imbres
> Quem super notas aluere ripas,
> Fervet, immensusque ruit—

but we must pause for a moment, lest we be hurried away by its tide. Let us endeavour calmly to consider whither it would carry us.

If by the elements returning *into themselves* be meant their return to higher levels, it is certainly possible that a fraction of the organic matter which is intermixed with the mud and sand deposited in alternate strata in the delta of the Ganges, may have been derived by the leaves and roots of plants from such aqueous vapour, carbonic acid, and other gases, as had ascended into the atmosphere from *lower* regions, and which were not, therefore, derived from the waste of rocks and their organic contents, or from the putrescence of vegetables previously nourished from these sources. This fraction, and this alone, may then be deducted from the mass of solid matter annually transported into the Bay of Bengal, and what re-

Address to the Geological Society on the Anniversary, Feb. 1831, p. 24.

mains, whether organic or inorganic, will be the measure of
the degradation which thousands of torrents in the Himalaya
mountains, and many rivers of other parts of India, bring
down in a single year.   Even in this case it will be found
that the sum of the force of vegetation can merely be considered as having been in a slight degree *conservative*, retard-
ing the waste of land, and not acting as an antagonist power.

But the untenable nature of the doctrine which we are now
controverting may be set in a clearer light by examining the
present state of the earth's surface, on which it is declared that
" an incalculable mass of solid matter is thrown down year
after year," in such a manner as to form a counterpoise to
the agents of decay.   Is it not a fact that the vegetable mould
is seldom more than a few feet in thickness, and that it often
does not exceed a few inches ?   Do we find that its volume
is more considerable on those parts of our continents which we
can prove, by geological data, to have been elevated at more
ancient periods, and where there has been the greatest time
for the accumulation of vegetable matter, produced throughout successive zoological epochs?   On the contrary, are not
these higher and older regions more frequently denuded, so as
to expose the bare rock to the action of the sun and air ?

Do we find in the torrid zone, where the growth of plants is
most rank and luxurious, that accessions of matter due to their
agency are most conspicuous on the surface of the land ?   On
the contrary, is it not there where the vegetation is most active
that, for reasons to be explained in the next chapter, even
those superficial peat mosses are unknown which cover a large
area in some parts of our temperate zone ?   If the operation
of animal and vegetable life could restore to the general surface of the continents a portion of the elements of those dis-
integrated rocks, of which such enormous masses are swept
down annually into the sea, along particular river-courses and
lines of coast, the effects would have become ere now most
striking ; and would have constituted one of the most leading
features in the structure and composition of our continents.

All the great steppes and table-lands of the world, where the action of running water is feeble, would have become the grand repositories of organic matter, accumulated without that intermixture of sediment which so generally characterizes the subaqueous strata.

Even the formation of peat in certain districts where the climate is cold and moist, the only case, perhaps, which affords the shadow of a support to the theory under consideration, has not in every instance a conservative tendency. A peat-moss often acts like a vast sponge, absorbing water in large quantities, and swelling to the height of many yards above the surrounding country. The turfy covering of the bog serves, like the skin of a bladder, to retain for a while the fluid within, and a violent inundation sometimes ensues when that skin bursts, as has often happened in Ireland, and many parts of the continent. Examples will be mentioned by us in a subsequent chapter, where the Stygian torrent has hollowed out ravines and borne along rocks and sand, in countries where such ravages could not have happened but for the existence of peat. Here, therefore, the force of vegetation accelerates the rate of decay of land, and the solid matter swept down to lower levels during such floods, counterbalances, to a certain degree, the accessions of vegetable mould which may accrue to the land by the growth of peat.

We may explain more clearly the kind of force which we imagine vegetation to exert, by comparing it to the action of frost, which augments the height of some few Alpine summits by causing a mass of perpetual snow to lodge thereon, or fills up some valleys with glaciers; but although by this process of congelation the rain-water that has risen by evaporation from the sea, is retained for awhile in a solid form upon the land, and although some elevated spots may be protected from waste by a constant covering of ice, yet by the sudden melting of snow and ice, the degradation of rocks is often accelerated. Although every year fresh snow and ice are formed, as also more vegetable and animal matter, yet there is no increase;

the one melts, the other putrifies, or is drifted down to the sea by rivers. If this were not the case, frost might be considered as an antagonist power, as well as the action of animal and vegetable life, and these by their combined energy might restore to continents a portion of that solid matter which is swept down into the sea from mountains and wasting cliffs. By the aid of such machinery might a theorist repair the losses of the solid land, sand and rocky fragments being carried down annually to the subaqueous regions from hills of granite, limestone, and shale, while vegetation and frost might raise new mountains, which, like the cliffs in Eschscholtz's Bay, might consist of icebergs, intermixed with vegetable mould.

We have stated in a former volume that, in the known operation of the *igneous* causes, a real antagonist power is found which may counterbalance the levelling action of running water; and there seems no good reason for presuming that the upheaving and depressing force of earthquakes, together with the heaping up of ejected matter by volcanos, may not be fully adequate to restore the superficial inequalities which rivers and oceanic currents annually tend to lessen. If a counterpoise be derived from this source, the quantity and elevation of land above the sea may for ever remain the same, in spite of the action of the aqueous causes, which, if thus counteracted, may never be able to reduce the surface of the earth more nearly to a state of equilibrium than that which it has now attained; and, on the other hand, the force of the aqueous agents themselves might thus continue for ever unimpaired. This permanence of the intensity of the powers now in operation would account for any amount of disturbance or degradation of the earth's crust, so far as the *mere quantity* of movement or decay is concerned; provided only that indefinite periods of time are contemplated.

As to the *intensity* of the disturbing causes at particular epochs, their effects have as yet been studied for too short a time to enable us fully to compare the signs of ancient convulsions with the permanent monuments left in the earth's crust by the events of the last few thousand years. But notwithstanding the small number of changes which have been witnessed and carefully recorded, observation has at least shown that our knowledge of the extent of the subterranean agency, as now developed from time to time, is in its infancy; and there can be no doubt that great partial mutations in the structure of the earth's crust are brought about in volcanic regions, without any interruption to the general tranquillity of the habitable surface.

Some geologists point to particular cases of enormous dislocation of ancient date, and confessedly not of frequent occurrence, where shifts in the strata of two thousand feet and upwards appear to have been produced suddenly and at one effort. But they have been at no pains to prove that similar consequences could not result from earthquakes such as have happened within the last three thousand years. They have usually proceeded on à priori reasoning to assume that such convulsions were paroxysmal, and attended by catastrophes such as have never occurred in modern times. It would be irrelevant to the subject immediately under consideration to enter into a long digression on these topics, but we may remind the reader, that the subsidence of the quay at Lisbon to the depth of six hundred feet only gave rise to a slight whirlpool ; and we may thence infer the possibility of a sinking down or elevation four or five times as great, especially in deeper seas, without any superficial disturbance unparalleled in the events of the last century.

If a certain sect of geologists were as anxious to reconcile the actual and former course of nature as they are eager to contrast them, they would perceive that the effects witnessed by us of subterranean action on supramarine land, may not be a type of those which the submerged rocks undergo, and they would proceed with more caution when reasoning from a comparison between the accumulated results of disturbing causes in the immensity of past time, and those which are recorded in the meagre annals of a brief portion of the human era.

The same rash generalizations which are now made respecting eras of paroxysmal violence and chaotic derangement, led formerly to the doctrines of universal formations, the improbability of which might have been foreseen by a slight reference to the causes now in operation.

To the same modes of philosophising we may ascribe the unwillingness of some naturalists to admit, that all the fossil species are not the same as those now living on the globe; whereas, if the facts and reasoning set forth in a former part of this volume, respecting the present instability of the organic creation be just, it might always à *priori* have been seen that the species inhabiting the planet at two periods very remote could hardly be identical.

In our view of the Huttonian theory, we pointed out as one of its principal defects, the assumed want of synchronism in the action of the great antagonist powers—the introduction, first, of periods when continents gradually wasted away, and then of others when new lands were elevated by violent convulsions. In order to have a clear conception of the working of such a system, let the reader suppose the earthquakes and volcanic eruptions of the Andes to be suspended for a million of years, and sedimentary deposits to accumulate throughout the whole of that period, as they now accumulate at the mouths of the Orinoco and Amazon, and along the intervening coast. Then let a period arrive when the subterranean power, which had obtained no vent during those ten thousand centuries, should escape suddenly in one tremendous explosion.

It is natural that geologists who reject such portions of the Huttonian theory as we embrace, should cling fondly to those parts which we deem unsound and unphilosophical. They have accordingly selected the distinctness of the periods when the antagonist forces are developed, as a principle peculiarly worthy of implicit faith. For this reason they have declined making any strenuous effort to account for those violations of continuity in the series of geological phenomena which are exhibited in large but limited regions; and which we have hinted may admit of explanation by the shifting of the volcanic foci, without the necessity of calling in to our aid any hypothetical eras of convulsion.

In the Oriental cosmogonies, as we have seen, both the physical and moral worlds were represented to be subject to gradual deterioration, until a crisis arrived when they were annihilated, or reverted to a state of chaos;—there had been alternating periods of tranquillity and disorder—an endless vicissitude of destructions and renovations of the globe.

In the spirit of this antique philosophy, some modern geologists conceive that nature, after long periods of repose, is agitated by fits of " feverish spasmodic energy, during which her very frame-work is torn asunder * ;"—these paroxysms of internal energy are accompanied by the sudden elevation of mountain chains, " followed by mighty waves desolating whole regions of the earth †"; and, according to some authors, whole races of organic beings are thus suddenly annihilated.

It was to be expected that when, in opposition to these favourite dogmas, we enumerated the subterranean catastrophes of the last one hundred and forty years, pointing out how defective were our annals, and called on geologists to multiply the amount of disturbances arising from this source by myriads of ages during the existence of successive races of organic beings, that we should provoke some vehement expostulation. We could not hope that the self-appointed guardians of Nature's slumber would allow us with impunity thus suddenly to intrude upon her rest, or that they would fail to resent so rude an attempt to rouse her from the torpor into which she had been lulled by their hypothesis. We were prepared to see our proofs and authorities severely sifted, our

* Prof. Sedgwick, Anniv. Address, &c. 1831, p. 35.   † Ibid.

inferences rigorously scrutinized; but we never supposed it possible that our adversaries would set up " as a vast counterpoise to all the agents of destruction," a cause so nugatory as " the single operation of vegetable life *."

As it will appear from what we before said, that vegetation cannot act as an antagonist power amid the mighty agents of change which are always modifying the surface of the globe, let us next inquire how far its influence is conservative,—how far it may retard the levelling power of running water, which it cannot oppose, much less counterbalance.

It is well known that a covering of herbage and shrubs may protect a loose soil from being carried away by rain, or even by the ordinary action of a river, and may prevent hills of loose sand from being blown away by the wind. For the roots bind together the separate particles into a firm mass, and the leaves intercept the rain-water, so that it dries up gradually instead of flowing off in a mass and with great velocity. The old Italian hydrographers make frequent mention of the increased degradation which has followed the clearing away of natural woods in several parts of Italy. A remarkable example was afforded in the Upper Val d'Arno, in Tuscany, on the removal of the woods clothing the steep declivities of the hills by which that valley is bounded. When the ancient forest laws were abolished by the Grand Duke Joseph, during the last century, a considerable tract of surface in the Cassentina (the Clausentinium of the Romans) was denuded, and, immediately, the quantity of sand and soil washed down into the Arno increased enormously. Frisi, alluding to such occurrences, observes, that as soon as the bushes and plants were removed, the waters flowed off more rapidly, and, in the manner of floods, swept away the vegetable soil †.

This effect of vegetation is of high interest to the geologist, when he is considering the formation of those valleys which have been principally due to the action of rivers. The spaces

* Prof. Sedgwick's Anniv. Address, Feb. 1831, p. 24.
† Treatise on Rivers and Torrents, p. 5, Garston's translation.

intervening between valleys, whether they be flat or ridgy, when covered with vegetation, may scarcely undergo the slightest waste, as the surface may be protected by the green sward of grass ; and this may be renewed, in the manner before described, from elements derived from rain-water and the atmosphere. Hence, while the river is continually bearing down matter in the alluvial plain, and undermining the cliffs on each side of every valley, the height of the intervening rising grounds may remain stationary.

In this manner a cone of loose scoriæ, sand and ashes, such as Monte Nuovo, may, when it has once become densely clothed with herbage and shrubs, suffer scarcely any farther dilapidation ; and the perfect state of the cones of hundreds of extinct volcanos in France, Campania, Sicily, and elsewhere, may prove nothing whatever, either as to their relative or absolute antiquity. We may be enabled to infer from the integrity of such conical hills of incoherent materials, that no flood can have passed over the countries where they are situated since their formation ; but the atmospheric action alone in spots where there happen to be no torrents, and where the surface was clothed with vegetation, could scarcely in any lapse of ages have destroyed them.

During a late tour in Spain I was surprized to see a district of gently undulating ground in Catalonia, consisting of red and grey sandstone, and in some parts of red marl, almost entirely denuded of herbage, while the roots of the pines, holm oaks, and some other trees were half exposed, as if the soil had been washed away by a flood. Such is the state of the forests, for example, between Orista and Vich, and near San Lorenzo. Being at length overtaken by a violent thunderstorm, in the month of August, I saw the whole surface, even the highest levels of some flat-topped hills, streaming with mud, while on every declivity the devastation of torrents was terrific. The peculiarities in the physiognomy of the district were at once explained, and I was taught that in speculating on the greater effects which the direct action of rain may once

have produced on the surface of certain parts of England, we
need not revert to periods when the heat of the climate was
tropical.

In the torrid zone the degradation of land is generally
more rapid, but the waste is by no means proportioned to
the superior quantity of rain or the suddenness of its fall,
the transporting power of water being counteracted by a
greater luxuriance of vegetation. A geologist who is no
stranger to tropical countries observes, that the softer rocks
would speedily be washed away in such regions, if the numerous roots of plants were not matted together in such a
manner as to produce considerable resistance to the destructive
power of the rains. The parasitical and creeping plants also
entwine in every possible direction so as to render the forests
nearly impervious, and the trees possess forms and leaves best
calculated to shoot off the heavy rains, which when they have
thus been broken in their fall are quickly absorbed by the
ground beneath, or when thrown into the drainage depressions
give rise to furious torrents *.

The felling of forests has been attended, in many countries,
by a diminution of rain, as in Barbadoes and Jamaica †. For
in tropical countries, where the quantity of aqueous vapour in
the atmosphere is very great, but where, on the other hand,
the direct rays of the sun have immense power, any impediment
to the free circulation of air, or any screen which shades the
earth from the solar rays, becomes a powerful cause of humidity, and wherever dampness and cold have begun to be
generated by such causes, the condensation of vapour continues. The leaves moreover of all plants are alembics, and
some of those in the torrid zone have a remarkable power of
distilling water, thus contributing to prevent the earth from
becoming parched up.

There can be no doubt that the state of the climate, especially the humidity of the atmosphere, influences vegetation,
and that, in its turn, vegetation reacts upon the climate;

* De la Beche, Geol. Man. p. 184.   † Phil. Trans., vol. ii., p. 294.

but some writers seem to have attributed too much import-ance to the influence of forests, particularly those of America, as if they were the primary cause of the moisture of the climate.

The theory of a modern author on this subject, "that forests exist in those parts of America only where the predominant winds carry with them a considerable quantity of moisture from the ocean," seems far more rational. In all countries, he says, " having a summer heat exceeding 70°, the presence or absence of natural woods, and their greater or less luxuriance, may be taken as a measure of the amount of humidity, and of the fertility of the soil. Short and heavy rains, in a warm country, will produce grass, which, having its roots near the surface, springs up in a few days, and withers when the moisture is exhausted ; but transitory rains, however heavy, will not nourish trees, because, after the surface is saturated with water, the rest runs off, and the moisture lodged in the soil neither sinks deep enough, nor is in sufficient quantity, to furnish the giants of the forest with the necessary sustenance. It may be assumed, that twenty inches of rain falling moderately, or at intervals, will leave a greater permanent supply in the soil than forty inches falling, as it sometimes does in the torrid zone, in as many hours *."

" In all regions," he continues, " where ranges of moun-tains intercept the course of the constant or predominant winds, the country on the windward side of the mountains will be moist, and that on the leeward dry, and hence parched deserts will generally be found on the west side of countries within the tropics, and on the east side of those beyond them, the prevailing winds in these cases being gene-rally in opposite directions. On this principle, the position of forests in North and South America may be explained. Thus, for example, in the region within the thirtieth parallel, the moisture swept up by the trade-wind from the Atlantic is pre-

---

* Maclaren, Art. America, Encyc. Britannica.

cipitated in part upon the mountains of Brazil, which are but low and so distributed as to extend far into the interior. The portion which remains is borne westward, and, losing a little as it proceeds, it is at length arrested by the Andes, where it falls down in showers on their summits. The aërial current, now deprived of all the humidity with which it can part, arrives in a state of complete exsiccation at Peru, where, consequently, no rain falls. In the same manner the Ghauts in India, a chain only three or four thousand feet high, intercept the whole moisture of the atmosphere, having copious rains on their windward side, while on the other the weather remains clear and dry. The rains in this case change regularly from the west side to the east, and vice versâ, *with the monsoons.* But in the region of America, beyond the thirtieth parallel, the Andes serve as a screen to intercept the moisture brought by the prevailing winds from the Pacific Ocean; rains are copious on their summits, and in Chili on their *western* declivities; but none falls on the plains to the *eastward,* except occasionally when the wind blows from the Atlantic*."

We have been more particular in explaining these views, because they appear to us to place in a true light the dependence of vegetation on climate, notwithstanding the reciprocal action which each exerts on the other, the humidity being increased, and more uniformly diffused throughout the year, by the gradual spreading of wood.

Before concluding this chapter, we must offer a few observations on the influence of man in modifying the physical geography of the globe, for we must class his agency among the powers of organic nature.

The modifications of the surface, resulting from human agency, are only on a considerable scale when we have obtained so much knowledge of the working of the laws of nature as to be able to use them as instruments to effect our purposes. We

---

* Maclaren, ibid., where the position of the American forests, in accordance with this theory, is laid down in a map.

must command nature by obeying her laws, according to the saying of the philosopher, and for this reason we can never materially interfere with any of the great changes which either the aqueous or igneous causes are bringing about on the earth. In vain would the inhabitants of Italy strive to prevent the tributaries of the Po and Adige from bearing down, annually, an immense volume of sand and mud from the Alps and Apennines; in vain would they toil to re-convey to the mountains the mass torn from them year by year, and deposited in the form of sediment in the Adriatic. But they have, nevertheless, been able to vary the distribution of this sediment over a considerable area, by embanking the rivers, and preventing the sand and mud from being spread, by annual inundations, over the plains.

We have explained how the form of the delta of the Po has been altered by this system of embankment, and how much more rapid, in consequence of these banks, have been the accessions of land at the mouths of the Po and Adige within the last twenty centuries. There is a limit, however, to these modifications, since the danger of floods augments with the increasing height of the river-beds, while the expense of maintaining the barrier is continually enhanced, as well as the difficulty of draining the low surrounding country.

In the Ganges, says Major R. H. Colebrooke, no sooner is a slight covering of soil observed on a new sand-bank, than the island is cultivated; water-melons, cucumbers, and mustard, become the produce of the first year, and rice is often seen growing near the water's edge, where the mud is in large quantity. Such islands may be swept away before they have acquired a sufficient degree of stability to resist permanently the force of the stream; but if, by repeated additions of soil, they acquire height and firmness, the natives take possession, and bring over their families, cattle and effects. They choose the highest spots for the sites of villages, where they erect their dwellings with as much confidence as they would do on the main land; for although the foundation is sandy, the uppermost soil being interwoven with the roots of grass and other plants, and hardened by the sun, is capable of withstanding all attacks of the river. These islands often grow to a considerable size, and endure for the lives of the new possessors, being only at last destroyed by the same gradual process of undermining and encroachment to which the banks of the Ganges are subject *.

If Bengal were inhabited by a nation more advanced in opulence and agricultural skill, they might, perhaps, succeed in defending these possessions against the ravages of the stream for much longer periods; but no human power could ever prevent the Ganges, or the Mississippi, from making and unmaking islands. By fortifying one spot against the set of the current, its force is only diverted against some other point; and, after a vast expense of time and labour, the property of individuals may be saved, but no addition would thus be made to the sum of productive land. It may be doubted, whether any system could be devised so conducive to *national* wealth, as the simple plan pursued by the peasants of Hindostan, who, wasting no strength in attempts to thwart one of the great operations of nature, permit the alluvial surface to be perpetually renovated, and find their losses in one place compensated in some other, so that they continue to reap an undiminished harvest from a virgin soil.

To the geologist, the Gangetic islands, and their migratory colonies, may present an epitome of the globe as tenanted by man. For during every century we cede some territory which the earthquake has sunk, or the volcano has covered by its fiery products, or which the ocean has devoured by its waves. On the other hand, we gain possession of new lands, which rivers, tides, or volcanic ejections have formed, or which subterranean causes have upheaved from the deep. Whether the human species will outlast the whole, or a great part of the continents and islands now seen above the waters, is a subject far beyond the reach of our conjectures; but thus much may be inferred

* Asiatic Trans., vol. vii.

from geological data,—that if such should be its lot, it will be no more than has already fallen to pre-existing species, some of which have, ere now, outlived the form and distribution of land and sea which prevailed at the era of their birth.

We have before shown, when treating of the excavation of new estuaries in Holland by inroads of the ocean, as also of the changes on our own coasts, that although the conversion of sea into land by artificial labours may be great, yet it must always be in subordination to the great movements of the tides and currents. If, in addition to the assistance obtained by parliamentary grants for defending Dunwich from the waves, all the resources of Europe had been directed to the same end, the existence of that port might possibly have been prolonged for many centuries. But, in the meantime, the current would have continued to sweep away portions from the adjoining cliffs on each side, rounding off the whole line of coast into its present form, until at length the town must have projected as a narrow promontory, becoming exposed to the irresistible fury of the waves.

It is scarcely necessary to observe, that the control which man can exert over the igneous agents is less even than that which he may obtain over the aqueous. He cannot modify the upheaving or depressing force of earthquakes, or the periods or degree of violence of volcanic eruptions; and on these causes the inequalities of the earth's surface, and, consequently, the shape of the sea and land, appear mainly to depend. The utmost that man can hope to effect in this respect, is occasionally to divert the course of a lava-stream, and to prevent the burning matter, for a season at least, from overwhelming a city, or other fruit of human industry.

No application, perhaps, of human skill and labour tends so greatly to vary the state of the habitable surface, as that employed in the drainage of lakes and marshes, since not only the *stations* of many animals and plants, but the general climate of a district, may thus be modified. It is also a kind of alteration to which it is difficult, if not impossible, to find anything analogous in the agency of inferior beings. For we ought always, before we decide that any part of the influence of man is novel and anomalous, carefully to consider all the powers of other animate agents which may be limited or superseded by him. Many who have reasoned on these subjects seem to have forgotten that the human race often succeeds to the discharge of functions previously fulfilled by other species; a topic on which we have already offered some hints, when explaining how the distribution and numbers of each species are dependent on the state of contemporary beings.

Suppose the growth of some of the larger terrestrial plants, or, in other words, the extent of forests, to be diminished by man, and the climate to be thereby modified, it does not follow that this kind of innovation is unprecedented. It is a change in the state of the vegetation, and such may often have been the result of the entrance of new species into the earth. The multiplication, for example, of certain insects in parts of Germany, during the last century, destroyed more trees than man, perhaps, could have felled during an equal period.

It is a curious fact, to which we shall again advert, that the sites of many European forests, cut down since the time of the Romans, have become peat-mosses; and thus a permanent change has been effected in these regions. But other woods, blown down by winds, in the same countries, have also become peat-bogs ; so that, although man may have accelerated somewhat the change, yet it may be doubted whether other animate and inanimate causes might not, without his interference, have produced similar results. The atmosphere of our latitudes may have been slowly and insensibly cooling down since the ancient forests began to grow, and the time may have arrived when slight accidents were sufficient to cause the decrease of trees, and the usurpation of their site by other plants.

We do not pretend to decide how far the power of man, to modify the surface, may differ in kind or degree from that of other living beings, but we suspect that the problem is more complex than has been imagined by many who have speculated

on such topics. If new land be raised from the sea, the greatest alteration in its physical condition, which could ever arise
from the influence of organic beings, would probably be produced by the first immigration of terrestrial plants, whereby the
tract would become covered with vegetation. The change next
in importance would seem to be when animals enter, and modify
the proportionate numbers of certain species of plants. If
there be any anomaly in the intervention of man, in farther
varying the relative numbers in the vegetable kingdom, it may
not so much consist in the kind or absolute quantity of alteration, as in the circumstance that *a single species*, in this case,
would exert, by its superior power and universal distribution,
an influence equal to that of hundreds of other terrestrial
animals.

If we inquire whether man, by his direct removing power,
or by the changes which he may give rise to indirectly, tends,
upon the whole, to lessen or increase the inequalities of the
earth's surface, we shall incline, perhaps, to the opinion that
he is a levelling agent. He conveys upwards a certain quantity of materials from the bowels of the earth in mining
operations ; but, on the other hand, much rock is taken
annually from the land, in the shape of ballast, and afterwards
thrown into the sea, whereby, in spite of prohibitory laws,
many harbours, in various parts of the world, have been blocked
up. We rarely transport heavy materials to higher levels,
and our pyramids and cities are chiefly constructed of stone
brought down from more elevated situations. By ploughing
up thousands of square miles, and exposing a surface for
part of the year to the action of the elements, we assist the
abrading force of rain, and destroy the conservative effects of
vegetation.

But the aggregate force exerted by man is truly insignificant,
when we consider the operations of the great physical causes,
whether aqueous or igneous, in the inanimate world. If all the
nations of the earth should attempt to quarry away the lava
which flowed during one eruption from the Icelandic volcanoes

in 1783 and the two following years, and should attempt to consign it to the deepest abysses of the ocean, wherein it might approach most nearly to the profundities from which it rose in the volcanic vent, they might toil for thousands of years before their task was accomplished. Yet the matter borne down by the Ganges and Burrampooter, in a single year, probably exceeds, in weight and volume, the mass of Icelandic lava produced by that great eruption.

CHAPTER XIII.

Effects produced by the action of animal and vegetable life on the material constituents of the earth's crust—Imbedding of organic remains in deposits on emerged land—Growth of Peat—Peat abundant in cold and humid climates—Site of many ancient forests in Europe now occupied by peat—Recent date of many of these changes—Sources of Bog Iron-ore—Preservation of animal substances in Peat—Causes of its antiseptic property—Miring of quadrupeds—Bursting of the Solway Moss—Bones of herbivorous quadrupeds found in peat—Imbedding of animal remains in Caves and Fissures—Formation of bony breccias—Human bones and pottery intermixed with the remains of extinct quadrupeds in caves in the South of France—Inferences deducible from such associations.

WE now come to the second subdivision of the inquiry explained in the preceding chapter,—the consideration of the permanent modifications produced in the material constituents of the earth's crust, by the action of animal and vegetable life.

New mineral compounds, such as might never have existed in this globe but for the action of the powers of vitality, are annually formed, and made to enter into deposits accumulated both above and beneath the waters. Although we can neither explain nor imitate the processes of animal and vegetable life whereby those substances are produced, yet we can investigate the laws by virtue of which organic matter becomes imbedded in new strata,—sometimes imparting to them a peculiar mineral composition,—sometimes leaving durable impressions and casts of the forms of animate beings in rocks, so as to modify their structure and appearance.

It has been well remarked by M. Constant Prevost, that the effects of geological causes are divisible into two great classes; those produced on the surface during the immersion of land beneath the waters, and those which take place after its emersion. Agreeably to this classification we shall consider, first, in what manner animal and vegetable remains become included and preserved in solid deposits on emerged land, or that part of the surface which is not *permanently* covered by water, whether of the sea or lakes; secondly, the manner in which organic remains become imbedded in sub-aqueous deposits.

Under the first division we shall treat of the following topics :—1st, the growth of peat, and the preservation of vegetable and animal remains therein;—2ndly, the preservation of animal remains in stalactite, and in the mud of caves and fissures;—3dly, the burying of organic remains in alluvium and the ruins of land-slips;—4thly, of the same in blown sand;—5thly, of the same in volcanic ejections, and alluvions composed of volcanic productions.

The growth of Peat and the preservation of Vegetable and Animal Remains therein.

THE generation of peat, when not completely under water, is confined to moist situations, where the temperature is low, and where vegetables may decompose without putrifying.   It may consist of any of the numerous plants which are capable of growing in such *stations :* but a species of moss (*sphagnum palustre*) constitutes a considerable part of the peat found in marshes of the north of Europe; this plant having the property of throwing up new shoots in its upper part, while its lower extremities are decaying *.   Reeds, rushes, and other aquatic plants may usually be traced in peat, and their organization is often so entire, that there is no difficulty in discriminating the distinct species.

In general, says Sir H. Davy, one hundred parts of dry peat contain from sixty to ninety-nine parts of matter destructible by fire, and the residuum consists of earths usually of the same kind as the substratum of clay, marl, gravel, or rock on which they are found, together with oxide of iron.   " The peat of the chalk counties of England," observes the same writer, " contains much gypsum; but I have found very little in any

---

* For a catalogue of the plants which contribute to the generation of peat, see Dr. Rennie on Peat, p. 171—178. Dr. Macculloch's Western Isles, vol. i. p. 129.

specimens from Ireland or Scotland, and in general these peats contain very little saline matter *.'' From the researches of Dr. Macculloch, it appears that peat is intermediate between simple vegetable matter and lignite, the conversion of peat to lignite being gradual, and being brought about in a great lapse of time by the prolonged action of water †.

Peat is sometimes formed on a declivity in mountainous regions where there is much moisture, but in such situations it rarely if ever exceeds four feet in thickness. In bogs, and in low grounds into which alluvial peat is drifted, it is found forty feet thick and upwards, but in such cases it generally owes one-half of its volume to the water which it contains. It nas seldom, if ever, been discovered within the tropics, and it rarely occurs in the valleys even in the south of France and Spain. It abounds more and more in proportion as we advance farther from the equator, and becomes not only more frequent but more inflammable in northern latitudes ‡; the cause of which may probably be that the carbonic acid and hydrogen, which are the most inflammable parts, do not readily assume the gaseous form in a cold atmosphere.

There is a vast extent of surface in Europe covered with peat, which in Ireland is said to extend over a tenth of the whole island. One of the mosses on the Shannon is described by Dr. Boates to be fifty miles long, by two or three broad; and the great marsh of Montoire, near the mouth of the Loire, is mentioned by Blavier as being more than fifty leagues in circumference. It is a curious and well-ascertained fact that many of these mosses of the north of Europe occupy the place of immense forests of pine and oak, which have many of them disappeared within the historical era. Such changes are brought about by the fall of trees and the stagnation of water, caused by their trunks and branches obstructing the free drainage of the atmospheric waters, and giving rise to a marsh. In a warm climate such decayed timber would immediately be

---

* Irish Bog Reports, p. 209.   † System of Geology, vol. ii. p. 353.
‡ Rev. Dr. Rennie, ibid. p. 260.

removed by insects, or by putrefaction; but, in the cold temperature now prevailing in our latitudes, many examples are recorded of marshes originating in this source. Thus, in Mar forest, in Aberdeenshire, large trunks of Scotch fir, which had fallen from age and decay, were soon immured in peat formed partly out of their perishing leaves and branches, and in part from the growth of other plants. We also learn that the overthrow of a forest by a storm, about the middle of the seventeenth century, gave rise to a peat moss, near Lochbroom, in Ross-shire, where, in less than half a century after the fall of the trees, the inhabitants dug peat *. Dr. Walker mentions a similar change when, in the year 1756, the whole wood of Drumlanrig was overset by the wind. Such events explain the occurrence, both in Britain and on the continent, of mosses where the trees are all broken within two or three feet of the original surface, and where their trunks all lie in the same direction †.

Nothing is more common than the occurrence of buried trees at the bottom of the Irish peat-mosses, as also in most of those of England, France, and Holland; and they have been so often observed with parts of their trunks standing erect, and with their roots fixed to the sub-soil, that no doubt can be entertained of their having generally grown on the spot. They consist for the most part of the fir, the oak, and the birch; where the sub-soil is clay, the remains of oak are the most abundant; where sand is the substratum, fir prevails. In the marsh of Curragh, in the Isle of Man, vast trees are discovered standing firm on their roots, though at the depth of eighteen or twenty feet below the surface. Some naturalists have desired to refer the imbedding of timber in peat mosses to aqueous transportation, since rivers are well known to float wood into lakes; but the facts above mentioned show that, in numerous instances, such an hypothesis is inadmissible. It has moreover been observed that in Scotland, as also in many parts of the continent, the largest trees are found in those peat mosses

* Dr. Rennie's Essays, p. 65.   † Ibid. p. 30.

which lie in the least elevated regions, and that the trees are proportionably smaller in those which lie at higher levels; from which fact De Luc and Walker have both inferred that the trees grew on the spot, for they would naturally attain a greater size in lower and warmer levels. The leaves also, and fruits of each species, are continually found immersed in the moss along with the parent trees, as, for example, the leaves and acorns of the oak, the cones and leaves of the fir, and the nuts of the hazel.

Sometimes, in the same bog, a stratification is observed of different kinds of wood, oak being found in the lowermost stratum, and birch and hazel in a second bed above. Sometimes still higher, a stratum, containing alder with the twigs of the bog myrtle (*Myrica galæ*), have been found *; the succession of strata, in this instance, indicating a gradual conversion of a dry tract into a swamp, and lastly a peat-moss.

The durability of pine-wood, which in the Scotch peat-mosses exceeds that of the birch and oak, is due to the great quantity of turpentine which it contains, and which is so abundant that the fir-wood from bogs is used by the country people, in parts of Scotland, in the place of candles. Such resinous plants, observes Dr. Macculloch, as fir, would produce a fatter coal than oak, because the resin itself is converted into bitumen †.

In Hatfield-moss, which appears clearly to have been a forest eighteen hundred years ago, fir-trees have been found ninety feet long, and sold for masts and keels of ships; oaks have also been discovered there above one hundred feet long. The dimensions of an oak from this moss are given in the Philosophical Transactions, No. 275, which must have been larger than any tree now existing in the British dominions.

In the same moss of Hatfield, as well as in that of Kincardine and several others, Roman roads have been found covered to the depth of eight feet by peat. All the coins, axes, arms, and other utensils found in British and French mosses, are also Roman; so that a considerable portion of the European peat-

---

* Lib. Ent. Know., Timber Trees, p. 32.   † Syst. of Geol., vol. ii., p. 356.

bogs are evidently not more ancient than the age of Julius
Cæsar. Nor can any vestiges of the ancient forests described
by that general, along the line of the great Roman way in
Britain, be discovered, except in the ruined trunks of trees in
peat.

De Luc ascertained that the very site of the aboriginal
forests of Hircinia, Semana, Ardennes, and several others, are
now occupied by mosses and fens; and a great part of these
changes have, with much probability, been attributed to the
strict orders given by Severus, and other emperors, to destroy
all the wood in the conquered provinces.   Several of the
British forests, however, which are now mosses, were cut at
different periods by order of the English parliament, because
they harboured wolves or outlaws.   Thus the Welsh woods
were cut and burnt in the reign of Edward I. ; as were many
of those in Ireland by Henry II., to prevent the natives from
harbouring in them and harassing his troops.

It is curious to reflect, that considerable tracts have by these
accidents been permanently sterilized, and that during a period
when civilization has been making great progress, large areas
in Europe have, by human agency, been rendered less capable
of administering to the wants of man.   Rennie observes with
truth, that in those regions alone which the Roman eagle
never reached—in the remote circles of the German empire, in
Poland and Prussia, and still more in Norway, Sweden, and
the vast empire of Russia—can we see what Europe was before
it yielded to the power of Rome *.   Desolation now reigns
where stately forests of pine and oak once flourished, such
as might now have supplied all the navies of Europe with
timber.

At the bottom of peat mosses there is sometimes found a cake,
or " pan," as it is termed, of oxide of iron, and the frequency of
bog iron-ore is familiar to the mineralogist.   The oak which
is so often found dyed black in peat, owes its colour to the
same metal.   From what source the iron is derived is by no

* Essays, &c., p. 74.

means obvious, since we cannot in all cases suppose that it has been precipitated from the waters of mineral springs. According to Fourcroy there is iron in all compact wood, and it is the cause of one-twelfth part of the weight of oak. The heaths (*Ericæ*) which flourish in a sandy, ferruginous soil, are said to contain more iron than any other vegetable.

It has been suggested that iron, being soluble in acids, may be diffused through the whole mass of vegetables, when they decay in a bog, and may, by its superior specific gravity, sink to the bottom, and be there precipitated, so as to form bog iron-ore; or where there is a subsoil of sand or gravel, it may cement them into ironstone or ferruginous conglomerate *.

One interesting circumstance attending the history of peat-mosses is the high state of preservation of animal substances buried in them for periods of many years. In June, 1747, the body of a woman was found six feet deep, in a peat-moor in the Isle of Axholm, in Lincolnshire. The antique sandals on her feet afforded evidence of her having been buried there for many ages; yet her nails, hair, and skin, are described as having shown hardly any marks of decay. In a turbary on the estate of the Earl of Moira, in Ireland, a human body was dug up, a foot deep in gravel, covered with eleven feet of moss; the body was completely clothed, and the garments seemed all to be made of hair. Before the use of wool was known in that country, the clothing of the inhabitants was made of hair, so that it would appear that this body had been buried at that early period; yet it was fresh and unimpaired †. In the Philosophical Transactions, we find an example recorded of the bodies of two persons having been buried in moist peat, in Derbyshire, in 1674, about a yard deep, which were examined twenty-eight years and nine months afterwards; " the colour of their skin was fair and natural, their flesh soft as that of persons newly dead ‡."

---

* Dr. Rennie, Essays, &c., p. 347.
† Ib. p. 521, where several other instances are referred to.
‡ Phil. Trans., vol. xxxviii., 1734.

Among other analogous facts we may mention, that in digging a pit for a well near Dulverton, in Somersetshire, many pigs were found in various postures, still entire. Their shape was well preserved, the skin, which retained the hair, having assumed a dry, membranous appearance. Their whole substance was converted into a white, friable, laminated, inodorous, and tasteless substance; but which, when exposed to heat, emitted an odour precisely similar to broiled bacon *.

We naturally ask whence peat derives this antiseptic property ? It has been attributed by some to the carbonic and gallic acids which issue from decayed wood, as also to the presence of charred wood in the lowest strata of many peat-mosses, for charcoal is a powerful antiseptic, and capable of purifying water already putrid. Vegetable gums and resins also may operate in the same way †.

The tannin occasionally present in peat is the produce, says Dr. Macculloch, of tormentilla, and some other plants, but the quantity he thinks too small, and its occurrence too casual, to give rise to effects of any importance. He hints that the soft parts of animal bodies, preserved in peat-bogs, may have been converted into adipocire by the action of water merely; an explanation which appears clearly applicable to some of the cases above enumerated ‡.

The manner, however, in which peat contributes to preserve, for indefinite periods, the harder parts of terrestrial animals, is a subject of more immediate interest to the geologist. There are two ways in which animals become occasionally buried in the peat of marshy grounds; they either sink down into the semifluid mud, underlying a turfy surface, upon which they have rashly ventured, or, at other times, a bog " bursts," in the manner described in a preceding chapter, and animals may be involved in the peaty alluvium.

In the extensive bogs of Newfoundland, cattle are sometimes found buried with their heads only and neck above ground,

* Dr. Rennie, Essays, &c., p. 521.   † Ibid., p. 531.
‡ Syst. of Geol., vol. ii., pp. 340—346.

and after having remained for days in this situation, they have been drawn out by ropes and saved. In Scotland, also, cattle venturing on the "quaking moss" are often mired, or "laired," as it is termed; and in Ireland, Mr. King asserts that the number of cattle which are lost in sloughs is quite incredible *.

The description given of the Solway moss will serve to illustrate the general character of these boggy grounds. That moss, observes Gilpin, is a flat area, about seven miles in circumference, situated on the confines of England and Scotland. Its surface is covered with grass and rushes, presenting a dry crust and a fair appearance; but it shakes under the least pressure, the bottom being unsound and semifluid. The adventurous passenger, therefore, who sometimes in dry seasons traverses this perilous waste, to save a few miles, picks his cautious way over the rushy tussocks as they appear before him, for here the soil is firmest. If his foot slip, or if he venture to desert this mark of security, it is possible he may never more be heard of.

"At the battle of Solway, in the time of Henry VIII. (1542), when the Scotch army, commanded by Oliver Sinclair, was routed, an unfortunate troop of horse, driven by their fears, plunged into this morass, which instantly closed upon them. The tale was traditional, but it is now authenticated; a man and horse, in complete armour, having been found by peat-diggers, in the place where it was always supposed the affair had happened. The skeleton of each was well preserved, and the different parts of the armour easily distinguished †."

This same moss, on the 16th of December, 1772, having been filled with water during heavy rains, rose to an unusual height and then burst. A stream of black half-consolidated mud began at first to creep over the plain, resembling, in the rate of its progress, an ordinary lava current. No lives were lost, but the deluge totally overwhelmed some cottages, and

* Phil. Trans., vol. xv., p. 949.
† Observations on Picturesque Beauty, &c., 1772.

covered four hundred acres. The highest parts of the original moss subsided to the depth of about twenty-five feet, and the height of the moss, on the lowest parts of the country which it invaded, was at least fifteen feet.

A recent inundation in Sligo (January, 1831) affords another example of this phenomenon. After a sudden thaw of snow the bog between Bloomfield and Geevah gave way, and a black deluge, carrying with it the contents of a hundred acres of bog, took the direction of a small stream, and rolled on with the violence of a torrent, sweeping along heath, timber, mud, and stones, and overwhelming many meadows and arable land. On passing through some boggy land the flood swept out a wide and deep ravine, and part of the road leading from Bloomfield to St. James's Well was completely carried away from below the foundation for the breadth of two hundred yards.

The antlers of large and full-grown stags are amongst the most common and conspicuous remains of animals in peat. They are not horns which have been shed, for portions of the skull are found attached, proving that the whole animal perished. Bones of the ox, hog, horse, sheep, and other herbivorous animals, also occur ; and in Ireland and the Isle of Man, skeletons of a gigantic elk ; but no remains have been met with belonging to those extinct quadrupeds of which the living congeners inhabit warmer latitudes, such as the elephant, rhinoceros, hippopotamus, hyæna, and tiger, though these are so common in superficial deposits of silt, mud, sand, or stalactite, in various localities throughout Great Britain. Their absence seems to imply that they had ceased to live before the atmosphere of this part of the world acquired that cold and humid character which favours the growth of peat.

From the facts before mentioned, that mosses occasionally burst, and descend in a fluid state to lower levels, it will readily be seen that lakes and arms of the sea may occasionally become the receptacles of drift-peat. Of this accordingly there are numerous examples, and hence the alternations of clay and sand with different deposits of peat so frequent on some coasts, as on those of the Baltic and German Ocean. We are informed by Deguer that remains of ships, nautical instruments, and oars, have been found in many of the Dutch mosses; and Gerard, in his History of the Valley of the Somme, mentions that in the lowest tier of that moss was found a boat loaded with bricks, proving that these mosses were at one period navigable lakes and arms of the sea, as were also many mosses on the coast of Picardy, Zealand, and Friesland, from which soda and salt are procured *. The canoes, stone hatchets, and stone arrow-heads, found in peat in different parts of Great Britain, lead to similar conclusions, —but these will more properly be considered when we treat of subaqueous phenomena.

Imbedding of Animal Remains in the Stalactite and Mud of Caves and Fissures.

WE explained in the former volume how vast fissures have been formed from time to time by earthquakes, and suggested that the continual percolation of acidulous waters through rocks of limestone might have enlarged these fissures into caverns. We shall now consider in what manner the remains of animals may become preserved in rents and cavities, confining ourselves at present to the monuments of events which are known or can be inferred to have happened within the human era.

As the same caves and fissures may remain open throughout periods of indefinite duration, and may become the receptacles of the remains of species inhabiting a country at very different epochs, it requires the utmost care to avoid confounding together the monuments of occurrences of very distinct dates. Dr. Buckland, in his indefatigable researches into this class of phenomena, has often guarded with great skill against such anachronisms, pointing out the comparatively recent preser-

* Rennie on Peat Moss, p. 205.

vation of some organic relics which have become mingled in a common tomb with those of older date.

Fissures are very common in calcareous rocks, and these are usually, in the course of ages, filled up in part by small angular fragments of limestone, which scale off under the influence of frost and rain. Vegetable earth and land-shells are washed in at the same time, and the whole mass often becomes cemented together by calcareous matter dissolved by rain-water, or supplied by mineral springs. In an uncultivated country the edges of such fissures are usually overgrown with bushes, so that herbivorous animals, especially when chased by beasts of prey, or when carelessly browzing on the shrubs, are liable to fall in and perish. Of this kind is a fissure still open in Duncombe Park, in Yorkshire, where the skeletons of dogs, sheep, goats, deer, and hogs, have been found, lodged upon different ledges that occur at various depths in a rent of the rock descending obliquely downwards*.

Above the village of Selside, near Ingleborough in Yorkshire, a chasm of enormous but unknown depth occurs in the scar-limestone, a member of the carboniferous series. "The chasm," says Professor Sedgwick, "is surrounded by grassy shelving banks, and many animals, tempted towards its brink, have fallen down and perished in it. The approach of cattle is now prevented by a strong lofty wall, but there can be no doubt that, during the last two or three thousand years, great masses of bony breccia must have accumulated in the lower parts of the great fissure, which probably descends through the whole thickness of the scar-limestone, to the depth of perhaps five or six hundred feet †."

A fissure in the limestone of the Coiron, in France, is seen on the high road between Aubenas and Ville-Dieu, filled with a breccia, consisting of angular fragments of the rock and land-shells cemented together. The mode of its formation is

---

* Buckland, Reliquiæ Diluvianæ, p. 55.

† Memoir on the Structure of the Lake Mountains of the North of England, &c., read before the Geological Society, January 5, 1831.

admirably illustrated by the rapid growth of a similar deposit not far distant. At the pass of Escrinet, on the northern escarpment of the Coiron hills, near Aubenas, a tabular mass of limestone is seen disintegrating into innumerable angular fragments, which are transported by the rain to the foot of the declivity, where they have accumulated at one spot, in a talus fifty feet in thickness and five hundred yards wide. The upper part is composed for the most part of loose fragments, on which numerous land-shells are seen living; the lower portion is consolidated by stalagmite into a compact mass which serves for mill-stones. The calcareous cement has a red tinge, but not of so deep a colour as most of the Mediterranean breccias *.

By the decomposition of the calcareous rocks near Nice, a soil is produced of a blood-red colour; and red breccias, consisting of fragments of rock and land-shells cemented together, are continually forming. If the mountains here were rent by earthquakes, we might expect the fissures to be gradually filled with red breccias like those of higher antiquity so celebrated in many parts of the Mediterranean.

It often happens that fissures communicate with subterranean caverns, a fact somewhat confirmatory of the views of those geologists who attribute the origin of limestone caverns in great part to the movements and dislocations of the strata. In this case the fissure may serve for ages as a natural pit-fall to animals passing by, and their bones, with all the earth, sand, and fragments of rock that fall through these passages, may be washed down or subside by their own weight, so as to reach the cavern below where thick deposits may be amassed.

Oftentimes when the bones of animals are strewed along the bottom of fissures or caves which they may have inhabited, they become covered over with mud introduced by land-floods, and are thus preserved from decomposition. Thus on the floor of many caverns mentioned by Dr. Buckland, in the Mendip Hills and Derbyshire, sedimentary mud has been left in recent times during floods.

* I examined this spot in the year 1828, accompanied by Mr. Murchison.

The same author observed in every cave examined by him in Germany, a deposit of mud or sand, sometimes with, and sometimes without, an intermixture of rolled pebbles and angular fragments of rock, and having its surface covered over with a *single* crust of stalagmite *. In the English caves he remarked a similar absence of *alternations* of alluvium and stalagmite. On the banks of the Meuse, however, at Chockier, about two leagues from Liège, a cavern has been recently discovered where there are three distinct beds of stalagmite, between each of which occur breccia and mud, mixed with some quartz pebbles, and the bones of extinct quadrupeds †.

But this exception does not invalidate the generality of the phenomenon observed by the Professor, and of which we have as yet seen no satisfactory explanation. The principal cause we suspect to be, that if several floods pass at different intervals of time through any subterranean passage, the last, if it has power to drift along fragments of rock, will also tear up any alternating stalagmitic and alluvial beds that may happen to have been previously formed. Another cause may be, that in a country in which torrents and rivers are gradually deepening their channels, and cutting through masses of cavernous limestone (an excavating process which is most rapid during epochs of subterranean disturbance, when the levels of a district are altered), it will only happen once that the stream will break into hollows or fissures communicating with a certain series of caverns. When the erosive action has proceeded farther, and the river has sunk to a greater depth, the drainage of the country will be effected in a valley at a level inferior to that of the caves, and consequently no transported matter will afterwards be introduced into them.

In the cave of Paviland, called Goat's Hole, on the coast of Glamorganshire, besides the bones of many extinct animals which occur in a mass of loam, a modern breccia has been formed, consisting of earth cemented by stalagmite, and con-

* Rel. Dil. p. 108.
† Journ. de Géologie, tome i. p. 286. July, 1830.

taining marine-shells and birds' bones, all of recent species.
The mouth of this cave is from thirty to forty feet above high-
water mark, being situated in a lofty cliff of limestone, facing
the estuary of the Severn, the waves of which, during great
storms, occasionally dash into it.   The left side of a human
skeleton was also found here under a cover of six inches of
earth.   In a cavernous aperture leading from the roof of this
cave to the face of the cliff was discovered a bed of brown
earth, apparently derived from dust driven in continually by the
wind ; and in this earth were the bones of various birds, of moles,
water-rats, mice, and fish, and a few land-shells, all clearly the
remains of modern animals.   Their presence in this almost
inaccessible spot can only be explained, says Dr. Buckland,
" by referring the bones of birds, moles, rats, and mice, to the
agency of hawks, and the fish-bones to that of sea-gulls.   The
land-shells, which are such as live at present on the rock without, may easily have fallen in.   Had there been any stalagmite uniting these bones into a breccia, they would have
afforded a perfect analogy to the accumulation of modern
birds' bones, by the agency of hawks, at Gibraltar *."

The formation last alluded to occurs in perpendicular fissures at the north extremity of the rock of Gibraltar, where a
reddish calcareous earth, containing numerous bones of small
birds, is in the act of accumulating.   Around these fissures a
number of hawks nestle and rear their young in the breeding-
season, and the bones are the remains of their food.   Major
Imrie mentions also a concretion in the rocks below King's
Lines, Gibraltar, consisting of pebbles of the prevailing calcareous rock, wherein, at a considerable depth under the surface, part of a green glass bottle was found imbedded †.

In a cave of mountain-limestone at Burringdon, in the
Mendip Hills, supposed to have been once used as a place of
sepulture or refuge, human bones have been met with, encrusted with stalactite, one of the skulls being filled with this

* Buckland, Reliquiæ Diluvianæ, p. 93.   † Ib. p. 156.

substance in the interior *. The state of the bones, says Dr. Buckland, affords indications of very high antiquity.

The remains of human skeletons have been also found in the cave of Wokey Hole, near Wells, in the Mendips, dispersed through reddish mud and clay, and some of them united by stalagmite into a firm osseous breccia. " The spot on which they lie is within reach of the highest floods of the adjacent river, and the mud in which they are buried is evidently fluviatile †."

We shall conclude with alluding to some caverns recently examined in the south of France, in which human bones and fragments of pottery are described as intermingled in the same deposits with the remains of extinct mammalia. We may first mention the cavern of Bize, in the department of Aude, where M. Marcel de Serres met with a small number of human bones mixed with those of extinct animals and with land-shells. They occur in a calcareous stony mass, bound together by a cement of stalagmite. On examining the same caverns, M. Tournal found not only in these calcareous beds, but also in a black mud which overlies a red osseous mud, several human teeth, together with broken angular fragments of a rude kind of pottery, and also marine and terrestrial shells of our own epoch. The teeth preserve their enamel, but the fangs are so much altered as to adhere strongly to the tongue. Of the terrestrial shells thus associated with the bones and pottery, the most common are Cyclostoma elegans, Bulimus decollatus, Helix nemoralis and H. nitida. Among the marine are found Pecten jacobæus, Mytilus edulis, and Natica mille-punctata, all of them eatable kinds. Bones of quadrupeds were found in the same mass belonging to three new species of the deer kind, an extinct bear (*Ursus arctoïdeus*), besides the wild bull (*Bos urus*), formerly a native of Germany‡.

In the same part of France, M. de Christol has found in

* Buckland, Reliquiæ, p. 164.   † Ib. p. 165.
‡ M. Marcel de Serres, Géognosie des Terrains Tertiaires, p. 64. Introduction.

caverns in a tertiary limestone at Pondres and Souvignargues, situated two leagues north of Lunel-viel, (department of Herault,) human bones and pottery confusedly mixed with the remains of the rhinoceros, bear, hyæna, and many other terrestrial mammifers. They were imbedded in an alluvial mud, of the solidity of calcareous tufa, and containing some flint pebbles and fragments of the limestone of the country. Beneath this mixed accumulation, which sometimes attained a thickness of thirteen feet, is the original floor of the cavern, about a foot thick, covered with bones and the dung of animals (*album græcum*), in a sandy and tufaceous cement.

The human bones in these caverns of Pondres and Souvignargues were found, upon a careful analysis, to have parted with their animal matter to as great a degree as those of the hyæna which accompany them, and are equally brittle, and adhere as strongly to the tongue.

In order to compare the degree of alteration of these bones with those known to be of high antiquity, M. Marcel de Serres, and M. Ballard, Chemist of Montpellier, procured some from a Gaulish sarcophagus in the plain of Lunel, supposed to have been buried for fourteen or fifteen centuries at least. In these the cellular tissue was empty, but they were more solid than fresh bones. They did not adhere to the tongue in the same manner as those of the caverns of Bize and Pondres, yet they had lost at least three-fourths of their original animal matter.

The superior solidity of the Gaulish bones to those in a fresh skeleton is a fact in perfect accordance with the observations made by Mr. Mantell on bones taken from a Saxon tumulus, near Lewes.

Let us now consider what conclusions are deducible from the important facts above enumerated. Must we infer that man and these extinct quadrupeds were contemporaneous inhabitants of the south of France at some former epoch? We should unquestionably have arrived at this conclusion if the bones had been found in an undisturbed *stratified* deposit of subaqueous origin, especially if it contained shells in regular layers like that of North-Cliff in Yorkshire, described by Mr. Vernon, from which we learn that the mammoth coexisted with thirteen species of our living British land and fresh-water testacea *. But we must hesitate before we draw analogous inferences from evidences so equivocal as that afforded by the mud, stalagmites and breccias of caves, where the signs of *successive* deposition are wanting.

No one will maintain that man, the hyæna, and the bear, were at once joint tenants of these caverns ; and if it be necessary to assume that the mud and pebbles were washed into their present position by floods, the same inundations might possibly have caught up the bones lying in more ancient deposits, and thus have mingled the whole together in the same mass.

More than ordinary caution is required in reasoning on the occurrence of human remains and works of art in alluvial deposits, since the chances of error are much greater than when we have the fossil bones of the inferior animals only under consideration. For the floor of caves has usually been disturbed by the aboriginal inhabitants of each country, who have used such retreats for dwelling places, or for concealment, or for sepulture. In such spots have treasures been often buried in periods of disturbance, or diligently sought for in times of tranquillity. The excavations made in Sicily for treasure-trove, in places where no money was ever buried, are believed to exceed in number all the spots in which coin was ever hid during the wars between the Saracens and Christians.

Dr. Buckland, in speaking of the cave of Paviland, before mentioned, states that the entire mass through which the bones were dispersed, appeared to have been disturbed by ancient diggings, so that the remains of extinct animals had, in that instance, actually become mixed with the recent bones and shells. In the same cave he found a human skeleton, and the remains of recent testacea of eatable species, which may have

---

* See ante, vol. i. p. 96.

been carried in by man. The same observation is applicable to the marine testacea of the cavern of Bize, and we suspect the whole phenomena to be very analogous.

To decide whether certain relics have been introduced by man, or natural causes, into masses of transported materials, must almost always be a task of some difficulty, especially where all the substances, organic and inorganic, have been mixed together and consolidated into one breccia; a change soon effected by the percolation of water charged with carbonate of lime. It is not on such evidence that we shall readily be induced to admit either the high antiquity of the human race, or the recent date of certain lost species of quadrupeds.

In continuing our investigation of the manner in which the animal and vegetable creation leave permanent marks of their existence on the *emerged* lands, we have next to examine,

The imbedding of organic remains in alluvium, and the ruins caused by landslips.

We restrict the term *alluvium* to such transported matter as has been thrown down, whether by rivers, floods, or other causes, upon land not *permanently* submerged beneath the waters of lakes or seas.

The alluvium of the bed of a river does not often contain any animal or vegetable remains, for the whole mass is so continually shifting its place, and the attrition of the various parts is so great, that even the hardest rocks contained in it are, at length, ground down to powder. But when sand, mud, and rubbish, are suddenly swept by a flood, and then let fall upon the land, such an alluvium may envelop trees or the remains of animals, which may, in this manner, be permanently preserved.

The sudden descent of a body of water which had been discharged by a small artificial drain from a lake in Vermont, in the United States, in 1810, covered a wide valley with the spoils of the land washed down from the higher country.

Deep accumulations of mud and sand were spread far and wide, and, in some places, deposits of timber were heaped up, from twenty to eighty feet in height *.

Analogous results happen, from time to time, when the course of a river has been obstructed by landslips, volcanic ejections, masses of ice, or other impediments, and when the waters of temporary lakes so caused burst through the barrier. Besides these indirect effects, the landslip, by suddenly precipitating large masses of rock and soil into a valley, overwhelms a multitude of animals, and sometimes buries permanently whole villages, with their inhabitants and large herds of cattle. Thus three villages, with their entire population, were covered, when the mountain of Piz fell in 1772, in the district of Treviso, in the state of Venice † ; and part of Mount Grenier, south of Chambery, in Savoy, which fell down in the year 1248, buried five parishes, including the town and church of St. André, the ruins occupying an extent of about nine square miles ‡.

The number of lives lost by the slide of the Rossberg, in Switzerland, in 1806, was estimated at more than eight hundred, a great number of the bodies being buried under mud and rock, at great depths, as well as several villages and scattered houses. In the same country, several hundred cottages, with eighteen of their inhabitants and a great number of cows, goats, and sheep, were victims to the sudden fall of a bed of stones, thirty yards deep, which descended from the summits of the Diablerets. In the year 1618, a portion of Mount Conto fell, in the county of Chiavenna in Switzerland, and buried the town of Pleurs with all its inhabitants, to the number of two thousand four hundred and thirty.

It is unnecessary to multiply examples of similar local catastrophes, which, however numerous they may nave been in the mountainous parts of Europe, within the historical period, have been, nevertheless, of rare occurrence in comparison to

* Ed. New Phil. Journ., No. III.,146.   † Malte-Brun's Geog., vol. i., 435.
‡ Bakewell, Travels in the Tarentaise, vol. i., p. 201.

the events of the same kind which take place in regions convulsed by earthquakes. It is then that all the causes whereby
terrestrial animals may be buried in superficial alluvium are
in full activity ; in proof of which, we need only refer the
reader to our description, in the former volume, of the effects
of great subterranean movements in disturbing the drainage of
a country and altering its levels. When the shocks are violent, enormous masses of rock and earth, even in comparatively low and level countries, are detached from the sides of
valleys and cast down into the river-courses. The slides are
so rapid and unexpected, that they often overwhelm, in the
day-time, every living thing upon the plain; and when they
happen in the night, escape is impossible. Although the
streams are often only partially dammed up by the ruins
thrown into their channel, the waters, nevertheless, collect in
sufficient quantity to form torrents of mud, which, as we have
seen in Calabria, sometimes bear along uprooted trees, and
overwhelm animals until, wherever they cease to move, the
mass shrinks on drying, and becomes hard and compact *.

Many geologists who seem desirous of ascribing as little
power as possible to the aqueous causes now acting, are in the
habit of overlooking the effects which the force of running
water can produce, when combined with the movements of
ordinary earthquakes. In a country like Great Britain, where
the height of mountain-chains is not considerable, and where
the shocks of earthquakes are rare and extremely feeble,
scarcely any remains of terrestrial animals or plants are buried
in alluvial deposits, in such a manner as to lead us to expect
that they will be preserved for indefinite periods. Some
skeletons, it is true, are occasionally imbedded, as, for example,
in the mud and sand produced by the floods in Scotland, in
1829, in which the dead and mutilated bodies of hares,
rabbits, moles, mice, partridges, and even the bodies of men,
were found drifted and partially buried †. But if the levels

* Vol. i., pp. 427 and 428.

† Sir T. D. Lauder, Bart., on the great floods in Morayshire, August, 1829, p. 177.

of a country remain unchanged, one flood usually effaces the memorials left by another, and there is rarely a sufficient depth of undisturbed transported matter in any one spot, to preserve the organic remains permanently from destruction.

The catastrophes, on the other hand, which arise from repeated earthquakes, cause not only the death of many animals, but their frequent inhumation in alluvium, so placed as to escape degradation for a succession of ages. When a valley has been half choked up with mud, sand, and gravel, or when numerous slides from the boundary hills have encumbered it with ruin, a river takes a new direction, finding, perhaps, its way through a new-formed fissure. From that moment the transported matter is no longer exposed to be undermined and removed by the action of running water.

Portions, also, of plains loaded with alluvial accumulations by transient floods, may be gradually upraised by earthquakes; and, if any organic remains have been imbedded in the transported materials, they will, after such elevation, remain undisturbed, and beyond the reach of the erosive power of streams. Every fissure, every hollow caused by the sinking in of land, becomes a receptacle of organic and inorganic substances, hurried along by transient floods, in districts where the drainage is repeatedly deranged by subterranean movements.

We have seen that the ravines which opened in Calabria, in 1783, were very numerous, varying in depth from fifty to two hundred feet *; and that animals were sometimes engulphed during the shocks. We may assume that many others fell in during the three years that the earthquakes continued, and that similar casualties happen in the intervals between convulsions. Every inundation, therefore, caused by heavy rains, every torrent that might chance to be in the line of any of these chasms, would pour in a quantity of mud, sand, and rolled pebbles, together with fragments of the adjacent rocks, and under these the animal remains might lie inhumed for ages.

Where houses with their inhabitants have been swallowed

* Vol. i., p. 421.

up in fissures, there appears to have been usually a sliding in
of all the loose matter which lay upon the surface, so that, in
such rents, we might look for rolled pebbles, mud, sand, angular fragments of rocks, ruins of buildings, and skeletons of men
and animals, at the depth often of several hundreds of feet.
It is obvious that a suite of subterranean caverns, communicating with such fissures, might become wholly, or partially,
filled with these various materials confusedly mixed together.

During the great earthquake of 1693, in Sicily, several
thousand people were at once entombed in the ruins of caverns
in limestone, at Sortino Vecchio; and, at the same time, a large
stream, which had issued for ages from one of the grottos
below that town, changed suddenly its subterranean course,
and came out from the mouth of a cave lower down the valley,
where no water had previously flowed.   To this new point the
ancient mills were transferred.

When the courses of engulphed rivers are thus liable to
change, from time to time, by alterations in the levels of a
country, and by the rending and shattering of mountain masses,
we must suppose that the dens of wild beasts will sometimes
be inundated by subterranean floods, and their carcasses buried
under heaps of alluvium.   The bones, moreover, of individuals which have died in the recesses of caves, or of animals
which have been carried in for prey, may be drifted along and
mixed up with mud, sand, and fragments of rock, so as to
form osseous breccias.

Nor should we omit to mention the havoc committed on low
coasts, during earthquakes, by waves of the sea which roll in
upon the land, bearing everything before them, for many
miles into the interior throwing down upon the surface great
heaps of sand and rock, by which the remains of drowned
animals may be overwhelmed.   Many of the storms, termed
hurricanes, have evidently been connected with submarine
earthquakes, as is shown by the atmospheric phenomena attend-
ant on them, and by the sounds heard in the ground, and the
odours emitted.   Such were the circumstances which accom-

panied the swell of the sea in Jamaica, in 1780, when a great wave desolated the western coast, and bursting upon Savanna la Mar, swept away the whole town in an instant, so that not a vestige of man, beast, or habitation, was seen upon the surface *.

Now let us suppose that in a tract of land constantly inhabited by terrestrial quadrupeds, the species are thrice changed under the gradual influence of causes before considered in this volume, and that, during the first and last of these zoological epochs, the district remains entirely free from earthquakes, but is violently convulsed by them during the intermediate era,— we should expect, for reasons above considered, that the fossil remains of quadrupeds, buried in alluvium, would be confined to one period only, viz., that of the subterranean movements. If the series of shocks should happen not to have occupied the whole of the second epoch, but only a small portion of it, there might be no indication whatever, in the fossil relics, of a passage from one state of the organic world to another. The transition would appear abrupt; and they who, for the sake of economizing past time, do not hesitate to magnify the energies of natural agents in by-gone ages, might then imagine one paroxysmal earthquake to have caused all the fissures, caverns, and depressions, and one accompanying deluge to have filled the whole with alluvial matter, annihilating, at the same time, the race of quadrupeds of which the bones remain interred.

But although we conceive that, in a country entirely free from subterranean movements, a long succession of ages might pass away without any permanent monuments being left in *alluvium* of the terrestrial animals which have lived upon the surface, yet it appears scarcely possible that man, if he has made considerable progress in civilization, should fail to leave some lasting signs of the works of his hands, to testify his former existence. We are informed by M. Boblaye, that in the Morea, the formation termed céramique, consisting of pottery, tiles, and bricks, intermixed with various works of art,

* Edwards, Hist. of West Indies, vol. i., p. 235, Ed. 1801.

enters so largely into the alluvium and vegetable soil upon the plains of Greece, and into hard and crystalline breccias which have been formed at the foot of declivities, that it constitutes a real stratum which might, in the absence of zoological characters, serve to mark our epoch in a most indestructible manner *.

Imbedding of Organic Bodies and Human Remains in Blown Sand.

The drifting of sand is the next cause which we may consider among those capable of preserving the remains of the inhabitants of the land during its period of emersion. We have already alluded to the African deserts, as the most remarkable example of desolation produced by this cause. Innumerable towns and cities have been buried to the westward of the Nile, between the temple of Jupiter Ammon and Nubia; and it is scarcely possible to conceive a mode whereby interment could take place under circumstances more favourable to the conservation of monuments for indefinite periods. The sand which surrounded and filled the great temple of Ipsambul, first discovered by Burckhardt, and afterwards partially uncovered by Belzoni and Beechey, was so fine as to resemble a fluid when put in motion. Neither the features of the colossal figures, nor the colour of the stucco with which some were covered, nor the paintings on the walls, had received any injury from being enveloped for ages in this dry impalpable dust †.

At some future period, perhaps, when the pyramids shall have perished, the action of the sea, or an earthquake, may lay open to the day some of these buried temples. Or we may suppose the desert to remain undisturbed, and changes in the surrounding sea and land to modify the climate and the direction of the prevailing winds, so that these may then waft away the Lybian sands as gradually as they once brought them to those regions. Thus many a town and temple of higher antiquity than Thebes or Memphis might reappear in their

* Ann. des Sci. Nat., tome xxii., p. 117. Feb. 1831.
† Stratton, Ed. Phil. Journ. No. V., p. 62.

original integrity, and a part of the gloom which overhangs the history of earlier nations might be dispelled.

Whole caravans are said to have been overwhelmed by the Lybian sands; and Burckhardt informs us that " after passing the Akaba, near the head of the Red Sea, the bones of dead camels are the only guides of the pilgrim through the wastes of sand." " We did not see," says Captain Lyon, speaking of a plain near the Soudah mountains, in Northern Africa, " the least appearance of vegetation; but observed many skeletons of animals, which had died of fatigue on the desert, and occasionally the grave of some human being. All these bodies were so dried by the heat of the sun, that putrefaction appears not to have taken place after death. In recently-expired animals I could not perceive the slightest offensive smell; and in those long dead the skin with the hair on it remained unbroken and perfect, although so brittle as to break with a slight blow. The sand-winds never cause these carcasses to change their places, for in a short time a slight mound is formed round them, and they become stationary *."

The burying of several towns and villages in England and France by blown sand is on record; thus for example in Suffolk, in the year 1688, part of Downham was overwhelmed by sands which had broken loose about one hundred years before, from a warren five miles to the south-west. This sand had, in the course of a century, travelled five miles, and covered more than a thousand acres of land †.

The ruins of buildings have been found entire in the drift-sand of Cornwall, as we before mentioned, as also land-shells. One of the latter is said to belong to a species which is unknown at present in this country ‡. Near St. Pol de Leon, in Brittany, a whole village was completely buried beneath drift-sand, so that nothing was seen but the spire of the church §.

---

* Travels in Northern Africa in the years 1818, 1819, and 1820, p. 83.
† Phil. Trans. vol. ii., p. 722.
‡ Vol. i., p. 301.
§ Mém. de l'Acad. des Sci. de Paris, 1772.—Malte-Brun's Geog. vol. i., p. 425.]

Imbedding of Organic Bodies and Works of Art in Volcanic Formations on the Land.

We have in some degree anticipated the subject of this section in a former volume, when speaking of the buried cities around Naples, and those on the flanks of Etna *. From the facts referred to by us, it appears that the preservation of human remains and works of art has been frequently due to the descent of floods caused by the copious rains which usually accompany eruptions. These aqueous lavas, as they are called in Campania, flow with great rapidity, and in 1822 surprised and suffocated, as we have stated, seven persons in the villages of St. Sebastian and Massa, on the flanks of Vesuvius.

In the tuffs, moreover, or solidified mud, deposited by these aqueous lavas, impressions of leaves and of trees have been observed. Some of those formed after the eruption of Vesuvius in 1822, are now preserved in the museum at Naples.

Lava itself may become indirectly the means of preserving terrestrial remains, by overflowing beds of ashes, pumice, and ejected matter, which may have been showered down upon animals and plants, or upon human remains. Few substances are better non-conductors of heat than volcanic dust and scoriæ, so that a bed of such materials is rarely melted by a superimposed lava-current. After consolidation, the lava affords secure protection to the lighter and more removeable mass below, wherein the organic relics may be enveloped. The Herculanean tuffs containing the rolls of papyrus, of which the characters are still legible, have, as we before remarked, been for ages covered by lava.

Another mode whereby lava may tend to the conservation of imbedded remains, at least of works of human art, is by overflowing them when not intensely heated, in which case they often suffer little or no injury.

Thus when the Etnean lava-current of 1669 covered fourteen towns and villages, and part of the city of Catania, it did not

* Vol. i., pp. 349 and 365.

melt down a great number of statues and other articles in the
vaults of Catania; and at the depth of thirty-five feet in the
same current, on the site of Mompiliere, one of the buried
towns, the bell of a church and some statues were found uninjured *.

We remarked in a former volume, that in many countries
which have been peopled from remote ages by civilized nations,
and have been at the same time the theatres of volcanic action,
there must be innumerable monuments of the highest value to
the historian, which continue unobserved " because they have
not been searched for." But we omitted to describe in detail
a splendid example of several buried cities in Central India,
which might probably be made to yield a richer harvest to the
antiquary than Pompeii and Herculaneum †. The city of
Oujein (or Oojain) was, about fifty years before the Christian
æra, the seat of empire, of art, and of learning; but in the time
of the Rajah Vicramaditya, it was overwhelmed, together, as
tradition reports, with more than eighty other large towns in
the provinces of Malwa and Bagur, " by a shower of earth."
The city which now bears the name is situated a mile to the
southward of the ancient town. On digging on the spot where
the latter is supposed to have stood, to the depth of fifteen or
eighteen feet, there are frequently discovered, says Mr. Hunter,
entire brick walls, pillars of stone, and pieces of wood of an
extraordinary hardness, besides utensils of various kinds, and
ancient coins. Many coins are also found in the channels cut
by the periodical rains, or in the beds of torrents into which
they have been washed. " During our stay at Oujein, a large
quantity of wheat was found by a man digging for bricks. It
was, as might have been expected, almost entirely consumed,
and in a state resembling charcoal. In a ravine cut by the
rains, from which several stone pillars had been dug, I saw a
space from twelve to fifteen feet long and seven or eight high,
composed of earthen vessels, broken and closely compacted
together. It was conjectured, with great appearance of probability, to have been a potter's kiln. Between this place and

* Vol. i., p. 366.   † Ibid., p. 407.

the new town is a hollow, in which, tradition says, the river
Sipparah formerly ran. It changed its course at the time the
city was buried, and now runs to the westward *." The soil
which covers Oujein is described as " being of an ash-grey
colour, with minute specks of black sand †."

That the " shower of earth " which is reported to have
" fallen from heaven," was produced by a volcanic eruption,
we cannot doubt, although no information has been obtained
respecting the site of the vent; and the nearest volcano of
which we read, is that which was in eruption during the Cutch
earthquake in 1819, at the distance of about thirty miles from
Bhooi, the capital of Cutch, and at least three hundred geographical miles from Oujein.

Captain F. Dangerfield, who accompanied Sir John Malcolm in his late expedition into Central India, states that the
river Nerbuddah, in Malwa, has its channel excavated through
columnar basalt, on which rest beds of marl impregnated with
salt. The upper of these beds is of a light colour, and from
thirty to forty feet thick, and rests horizontally on the lower
bed, which is of a reddish colour. Both appear from the
description to be tuffs composed of the materials of volcanic
ejections, and forming a covering from sixty to seventy feet
deep overlying the basalt, which seems to resemble some of the
currents of prismatic lava in Auvergne and the Vivarais. Near
the middle of this tufaceous mass, and therefore at the depth
of thirty feet or more from the surface, just where the two
beds of tuff meet, Captain Dangerfield was shown, near the
city of Mhysir, buried bricks and large earthen vessels, said to
have belonged to the ancient city of Mhysir, destroyed by the
catastrophe of Oujein ‡.

---

* Narrative of a Journey from Agra to Oujein, Asiatic Researches, vol. vi. p. 36.
          † Asiatic Journal, vol. ix. p. 35.
   ‡ Sir J. Malcolm's Cent. Ind.—Geol. of Malwa, by Captain F. Dangerfield,
App. No. ii. pp. 324, 325.

CHAPTER XV.

Imbedding of organic remains in subaqueous deposits—Division of the subject—Phenomena relating to terrestrial animals and plants first considered—Wood sunk to a great depth in the sea instantly impregnated with salt-water—Experiments of Scoresby—Drift timber carried by the Mackenzie into Slave Lake and into the sea—Cause of the abundance of drift timber in this river—Floating trees in the Mississippi—In the Gulf stream—Immense quantity thrown upon the coast of Iceland, Spitzbergen, and Labrador—Imbedding of the remains of insects—of the remains of reptiles—Why the bones of birds are so rare in subaqueous deposits—Imbedding of terrestrial quadrupeds—Effects of a flood in the Solway Firth—Wild horses annually drowned in the savannahs of South America—Skeletons in recent shell marl—Drifting of mammiferous and other remains by tides and currents.

We have treated hitherto of the imbedding of organic remains in deposits formed upon the emerged land, and we shall next consider the including of the same in deposits formed under water.

It will be convenient to divide this branch of our subject into three parts; considering first, the various modes whereby the relics of *terrestrial* species may be buried in subaqueous formations; secondly, the modes whereby the animals and plants inhabiting *fresh-water* may be so entombed; thirdly, the manner in which *marine* species may become preserved in new strata.

The phenomena which we are now about to notice demand a fuller share of attention than those previously examined, since the deposits which originate upon the dry land are insignificant in thickness, superficial extent, and durability, when contrasted with those of subaqueous origin. At the same time, the study of the latter is beset with greater difficulties, for we are here concerned with the results of processes much more removed from the sphere of ordinary observation. There is, indeed, no circumstance, as we before remarked *,

* Vol. i., p. 81.

which more seriously impedes the acquisition of just views in the etiology of our science, than an habitual disregard of the important fact, that the reproductive effects of the principal agents of change are confined to another element,—to that larger portion of the habitable globe, from which, by our very organization, we are almost entirely excluded.

Imbedding of Terrestrial Plants.

When a tree falls into a river from the undermining of the banks, or from being washed in by a torrent or flood, it floats on the surface, not because the woody portion is specifically lighter than water, but because it is full of pores containing air. When soaked for a considerable time, the water makes its way into these pores, and the wood becomes *water-logged* and sinks. The time required for this process varies differently in different woods, but several kinds may be drifted to great distances, sometimes across the ocean, before they lose their buoyancy.

If wood be sunk to vast depths in the sea, it may be impregnated with water suddenly. Captain Scoresby informs us, in his Account of the Arctic Regions *, that on one occasion a whale, on being harpooned, ran out all the lines in the boat, which it then dragged under water, the men having just time to escape to a piece of ice. When the fish returned to the surface " to blow," it was struck a second time, and soon afterwards killed. The moment it expired it began to sink,— an unusual circumstance, which was found to be caused by the weight of the sunken boat which still remained attached to it. By means of harpoons and ropes the fish was prevented from sinking until it was released from the weight by connecting a rope to the lines of the attached boat, which was no sooner done than the fish rose again to the surface. The sunken boat was then hauled up with great labour, for so heavy was it, that although before the accident it would have been buoyant when full of water, yet it now required a boat at each

* Vol. ii. p. 191.

end to keep it from sinking. " When it was hoisted into the ship, the paint came off the wood in large sheets; and the planks, which were of wainscot, were as completely soaked in every pore as if they had lain at the bottom of the sea since the Flood ! A wooden apparatus that accompanied the boat in its progress through the deep, consisting chiefly of a piece of thick deal, about fifteen inches square, happened to fall overboard, and, though it originally consisted of the lightest fir, sank in the water like a stone. The boat was rendered useless; even the wood of which it was built, on being offered to the cook for fuel, was tried and rejected as incombustible *."

Captain Scoresby found that by sinking pieces of fir, elm, ash, &c., to the depth of four thousand and sometimes six thousand feet, they became impregnated with sea-water, and when drawn up again, after immersion for an hour, would no longer float. The effect of this impregnation was to increase the dimensions as well as the specific gravity of the wood, every solid inch having increased one-twentieth in size and twenty-one twenty-fifths in weight †.

When timber is drifted down by a river, it is often arrested by lakes, and becoming water-logged it may sink and be imbedded in lacustrine strata, if any be there forming : sometimes a portion floats on till it reaches the sea. In the course of the Mackenzie River we have an example of vast accumulations of vegetable matter now in progress under both these circumstances.

In Slave Lake in particular,which vies in dimensions with some of the great fresh-water seas of Canada, the quantity of drift-timber brought down annually is enormous. " As the trees," says Dr. Richardson, " retain their roots, which are often loaded with earth and stones, they readily sink, especially when water-soaked, and, accumulating in the eddies, form shoals, which ultimately augment into islands. A thicket of small willows covers the new-formed island as soon as it appears above water, and their fibrous roots serve to bind the

* Account of the Arctic Regions, vol. ii. p. 193.   † Ib. p. 202.

whole firmly together. Sections of these islands are annually made by the river, assisted by the frost; and it is interesting to study the diversity of appearances they present according to their different ages. The trunks of the trees gradually decay until they are converted into a blackish brown substance resembling peat, but which still retains more or less of the fibrous structure of the wood; and layers of this often alternate with layers of clay and sand, the whole being penetrated, to the depth of four or five yards or more, by the long fibrous roots of the willows. A deposition of this kind, with the aid of a little infiltration of bituminous matter, would produce an excellent imitation of coal, with vegetable impressions of the willow roots. What appeared most remarkable was the horizontal slaty structure that the older alluvial banks presented, or the *regular curve* that the strata assumed from unequal subsidence.

" It was in the rivers only that we could observe sections of these deposits, but the same operation goes on on a much more magnificent scale in the lakes. A shoal of many miles in extent is formed on the south side of Athabasca Lake, by the drift-timber and vegetable debris brought down by the Elk River; and the Slave Lake itself must in process of time be filled up by the matters daily conveyed into it from Slave-River. Vast quantities of drift timber are buried under the sand at the mouth of the river, and enormous piles of it are accumulated on the shores of every part of the lake *."

The banks of the Mackenzie display almost everywhere horizontal beds of wood coal, alternating with bituminous clay, gravel, sand, and friable sandstone; sections, in short, of such deposits as are now evidently forming at the bottom of the lakes which it traverses.

Notwithstanding the vast forests intercepted by the lakes, a still greater mass of drift-wood is found where the Mackenzie reaches the sea, in a latitude where no wood grows at present except a few stunted willows. At the mouths of the river the

* Dr. Richardson's Geognost. Obs. on Capt. Franklin's Polar Expedition.

alluvial matter has formed a barrier of islands and shoals, where we may expect a great formation of coal at some distant period.

The abundance of floating timber on the Mackenzie is owing, as I am informed by Dr. Richardson, to the peculiar direction and to the length of the course of this river, which runs from south to north, so that the sources of the stream lie in much warmer latitudes than its mouths. In the country, therefore, where the former are situated, the frost breaks up at an earlier season, while yet the waters in the lower part of its course are ice-bound. Hence the current of water, rushing down northward, reaches a point where the thaw has not begun, and finding the channel of the river blocked up with ice, it overflows the banks, sweeping through forests of pines, and carrying away thousands of uprooted trees.

We have already observed * that the navigation of the Mississippi is much impeded by trunks of trees half sunk in the river. On reaching the Gulf of Mexico many of them subside and are imbedded in the new strata which form the delta, but many of them float on and enter the Gulf-stream. " Tropical plants, (says M. Constant Prevost,) are taken up by this great current, and carried in a northerly direction, till they reach the shores of Iceland and Spitzbergen uninjured. A great portion of them are doubtless arrested on their passage, and, probably, always in the same inlets, or the same spots on the bottom of the ocean ; in fact, wherever an eddy or calm determines their distribution, which, in this single example, extends over a space comprehended between the equator and the eightieth degree of latitude—an immense space, six times more considerable than that occupied by all Europe, and thirty times larger than France. The drifting of various substances, though regular, is not continual ; it takes place by intermittance after great inundations of rivers, and in the intervals the waters may only carry sand or mud, or each of these alternately, to the same localities †."

* Vol. i. p. 245.   † Mém. de la Soc. d'Hist. Nat. de Paris, vol. iv. p. 84.

The ancient forests of Iceland, as Malte-Brun observes, have been improvidently exhausted; but, although the Icelander can obtain no timber from the land, he is supplied with it abundantly by the ocean. An immense quantity of thick trunks of pines, firs, and other trees, are thrown upon the northern coast of the island, especially upon North Cape and Cape Langaness, and are then carried by the waves along these two promontories to other parts of the coast, so as to afford sufficiency of wood for fuel and for constructing boats. Timber is also carried to the shores of Labrador and Greenland; and Crantz assures us that the masses of floating wood thrown by the waves upon the island of John de Mayen often equal the whole of that island in extent*.

In a similar manner the bays of Spitzbergen are filled with drift-wood, which accumulates also upon those parts of the coast of Siberia that are exposed to the east, consisting of larch trees, pines, Siberian cedars, firs, and Fernambucco and Campeachy woods. These trunks appear to have been swept away by the great rivers of Asia and America. Some of them are brought from the Gulf of Mexico, by the Bahama stream, while others are hurried forward by the current which, to the north of Siberia, constantly sets in from east to west. Some of these trees have been deprived of their bark by friction, but are in such a state of preservation as to form excellent building timber †. Parts of the branches and almost all the roots remain fixed to the pines which have been drifted into the North Sea, into latitudes too cold for the growth of such timber, but the trunks are usually barked.

The leaves and lighter parts of plants are seldom carried out to sea, in any part of the globe, except during tropical hurricanes among islands, and during the agitations of the atmosphere which sometimes accompany earthquakes and volcanic eruptions ‡.

* Malte-Brun, Geog. vol. v. part i. p. 112.—Crantz, Hist. of Greenland, tome i. pp. 50—54.

† Olafsen, Voyage to Iceland, tome i.   Malte-Brun's Geog. vol. v. part i. p. 112.

‡ De la Beche, Geol. Manual, p. 477.

It will appear from these observations, that although the remains of terrestrial vegetation, borne down by aqueous causes from the land, are chiefly deposited at the bottom of lakes or at the mouths of rivers, yet a considerable quantity is drifted about in all directions by currents, and may become imbedded in any *marine* formation, or may sink down, when water-logged, to the bottom of unfathomable abysses, and there accumulate without intermixture of other substances.

It may be asked whether we have any data for inferring that the remains of a considerable proportion of the existing species of plants will be permanently preserved, so as to be hereafter recognizable, supposing the strata now in progress to be at some future period upraised? To this inquiry we may reply that there are no reasons for expecting that more than a small number of the plants now flourishing in the globe will become fossilized, since the entire habitations of a great number of them are remote from lakes and seas, and even where they grow near to large bodies of water, the circumstances are quite accidental and partial which favour the imbedding and con-servation of vegetable remains. Those naturalists, therefore, who infer that the ancient flora of the globe was, at certain periods, less varied than now, merely because they have as yet discovered only a few hundred fossil species of a particular epoch, while they can enumerate more than fifty thousand living ones, are reasoning on a false basis, and their standard of com-parison is not the same in the two cases.

Imbedding of the Remains of Insects.

I HAVE observed the elytra and other parts of beetles in a band of fissile clay, separating two beds of recent shell-marl, in the Loch of Kinnordy. Amongst these, Mr. Curtis recognized Elater lineatus and Atopa cervina, species still living in Scot-land. These, as well as other remains which accompanied them, appear to belong to terrestrial, not aquatic species, and must have been carried down in muddy water during an inundation.

In the lacustrine peat of the same locality, the elytra of beetles are not uncommon; but in the deposits of drained lakes generally, and in the silt of our estuaries, the relics of this class of the animal kingdom are extremely rare. In the blue clay of very modern origin of Lewes Levels, Mr. Mantell has found the Indusia, or cases of the larvæ of Phryganea, in abundance, with minute shells belonging to the genera Planorbis, Limnea, &c., adhering to them *.

When speaking of the migrations of insects, we pointed out that an immense number are floated into lakes and seas by rivers, or blown by winds far from the land; but they are so buoyant that we can only suppose them, under very peculiar circumstances, to sink to the bottom before they are either devoured by insectivorous animals or are decomposed.

Remains of Reptiles.

As the bodies of several crocodiles were found in the mud brought down to the sea by the river inundation which attended an earthquake in Java in the year 1699, we may imagine that extraordinary floods of mud may stifle many individuals of the shoals of alligators and other reptiles which frequent lakes and the deltas of rivers in tropical climates. Thousands of frogs were found leaping about among the wreck carried into the sea by the late inundations in Morayshire †; and it is evident that whenever a sea-cliff is undermined, or land is swept by other violent causes into the sea, land reptiles may be carried in.

Remains of Birds.

WE might have anticipated that the imbedding of the remains of birds in new strata would be of very rare occurrence, for their powers of flight insure them against perishing by numerous casualties to which quadrupeds are exposed during

---

* Trans. Geol. Soc. vol. iii. part i. p. 201, Second Series.
† Sir T. D. Lauder's Account, Second Ed., p. 312.

floods; and if they chance to be drowned, or to die when
swimming on the water, it will scarcely ever happen that they
will be submerged so as to become preserved in sedimentary
deposits. For in consequence of the hollow tubular structure
of their bones and the quantity of their feathers, they are
extremely light in proportion to their volume, so that when
first killed they do not sink to the bottom like quadrupeds,
but float on the surface until the carcass either rots away
or is devoured by predaceous animals. To these causes we
may ascribe the absence of any vestige of the bones of birds
in the recent marl formations of Scotland ; although these
lakes, until the moment when they were artificially drained,
were frequented by a great abundance of water-fowl.

Imbedding of Terrestrial Quadrupeds.

RIVER inundations recur in most climates at very irregular
intervals, and expend their fury on those rich alluvial plains
where herds of herbivorous quadrupeds congregate together.
These animals are often surprised, and being unable to stem the
current, are hurried along until they are drowned, when they
sink immediately to the bottom. Here their bodies are drifted
along, together with sediment, into lakes or seas, and may then
be covered by a mass of mud, sand, and pebbles, thrown down
upon them. If there be no sediment superimposed, the gases
generated by putrefaction usually cause the bodies to rise
again to the surface about the ninth, or at most the fourteenth
day. The pressure of a thin covering would not be sufficient
to retain them at the bottom, for we see the putrid carcasses
of dogs and cats, even in rivers, floating with considerable
weights attached to them, and they would be still more buoyant
in sea-water.

In cases where the body is so buried in drift-sand, or mud
accumulated upon it, as never to rise again, the skeleton
may be preserved entire; but if it comes again to the surface
while in the process of putrefaction, the bones commonly fall

piecemeal from the floating carcass, and may in that case become scattered at random over the bottom of a lake, estuary, or sea, so that a jaw may afterwards be found in one place, a rib in another, a humerus in a third—all included, perhaps, in a matrix of fine materials, and where there may be evidence of very slight transporting power in the current, or even of none, but simply of some chemical precipitate.

A large number of the bodies of drowned animals, if they float into the sea or a lake, especially in hot climates, are instantly devoured by sharks, alligators, and other carnivorous beasts, which may have power to digest even the bones. But during extraordinary floods, when the greatest number of land animals are destroyed, the waters are commonly so turbid, especially at the bottom of the channel, that even aquatic species are compelled to escape into some retreat where there is clearer water, lest they should be stifled. For this reason, as well as the rapidity of sedimentary deposition at such seasons, the probability of some carcasses becoming permanently imbedded is considerable.

One of the most memorable floods of modern date, in our island, is that which visited part of the southern borders of Scotland, on the 24th of January, 1794, and which spread particular devastation over the country adjoining the Solway Frith.

We learn from the account of Captain Napier, that the heavy rains had swollen every stream which entered the Frith of Solway, so that the inundation not only carried away a great number of cattle and sheep, but many of the herdsmen and shepherds, washing down their bodies into the estuary. After the storm, when the flood subsided, an extraordinary spectacle was seen on a large sand-bank, called " the beds of Esk," where there is a meeting of the tidal waters, and where heavy bodies are usually left stranded after great floods. On this single bank were found collected together the bodies of nine black cattle, three horses, one thousand eight hundred and forty sheep, forty-five dogs, one hundred and eighty hares, besides a great number of smaller animals, and, mingled with the rest, the corpses of two men and one woman *.

In those more recent floods in Scotland, in August 1829, whereby a fertile district, six hundred miles in length, became a scene of dreadful desolation, a vast number of animals and plants were washed from the land, and found scattered about after the storm, around the mouths of the principal rivers. An eye-witness thus describes the scene which presented itself at the mouth of the Spey, in Morayshire. " For several miles along the beach, crowds were employed in endeavouring to save the wood and other wreck with which the heavy rolling tide was loaded ; whilst the margin of the sea was strewed with the carcasses of domestic animals, and with millions of dead hares and rabbits. Thousands of living frogs, also, swept from the fields, no one can say how far off, were observed leaping among the wreck †."

We are informed by Humboldt, that during the periodical swellings of the large rivers in South America, great numbers of quadrupeds are annually drowned. Of the wild horses, for example, which graze in immense troops in the savannahs, thousands are said to perish when the river Apure is swollen, before they have time to reach the rising grounds of the Llanos. The mares, during the season of high water, may be seen, followed by their colts, swimming about and feeding on the grass of which the top alone waves above the waters. In this state they are pursued by crocodiles; and their thighs frequently bear the prints of the teeth of these carnivorous reptiles. " Such is the pliability," observes the celebrated traveller, " of the organization of the animals which man has subjected to his sway, that horses, cows, and other species of European origin, lead, for a time, an amphibious life, surrounded by crocodiles, water-serpents, and manatees. When the rivers return again into their beds, they roam in the savan-

* Treatise on Practical Store Farming, p. 25.
† Sir T. D. Lauder's Account of the Great Floods in Morayshire, August 1829, p. 312, Second Ed.

nah, which is then spread over with a fine odoriferous grass,
and enjoy, as in their native climate, the renewed vegetation
of spring *."

We find it continually stated, by those who describe the
Ganges and Burrampooter, that these rivers carry before them,
during the flood season, not only floats of reeds and timber,
but dead bodies of men, deer, and oxen †.

We have already referred to the effects of a flood which
attended an earthquake in Java in 1699, when the turbid
waters of the Batavian river destroyed all the fish except the
carp; and when drowned buffaloes, tigers, rhinoceroses, deer,
apes, and other wild beasts, were brought down to the sea-
coast by the current, with several crocodiles which had been
stifled in the mud ‡.

On the western side of the same island, in the territory of
Goulongong, in the regencies, a more recent volcanic eruption
(1821) was attended by a flood, during which the river Tjetan-
doy bore down hundreds of carcasses of rhinoceroses and
buffaloes, and swept away more than one hundred men and
women from a multitude assembled on its banks to celebrate
a festival. Whether the bodies reached the sea, or were deposited, with drift matter, in some of the large intervening
alluvial plains, we are not informed §.

We might enumerate a great number of local deluges that
have swept through the fertile lands which border on large
rivers, especially in tropical countries, but we should surpass
the limits of this work. We may observe, however, that the
destruction of islands, in rivers, is often attended with great
loss of lives. Thus, when the principal river in Virginia rose,
in 1771, to the height of twenty-five feet above its ordinary
level, it swept entirely away Elk Island, on which were seven

---

* Humboldt's Pers. Narr., vol. iv., pp. 394—396.
† Malte-Brun, Geog., vol. iii., p. 22.   ‡ See ante, vol. i., p. 444.
§ This account I had from Mr. Baumhauer, Director-General of Finances in
Java.

hundred head of quadrupeds,—horses, oxen, sheep, and hogs,—and nearly one hundred houses *.

The reader will gather, from what we have said in a former volume respecting the deposition of sediment by aqueous causes, that the greater number of the remains of quadrupeds drifted away by rivers must be intercepted by lakes before they reach the sea, or buried in fresh-water formations near the mouths of rivers. If they are carried still farther, the probabilities are increased of their rising to the surface in a state of putrefaction, and, in that case, of being there devoured by aquatic beasts of prey, or of subsiding into some spots whither no sediment is conveyed, and, consequently, where every vestige of them will, in the course of time, disappear.

In some instances, the skeletons of quadrupeds are met with abundantly in recent shell-marls in Scotland, where we cannot suppose them to have been imbedded by the action of rivers or floods. They all belong to species which now inhabit, or are known to have been indigenous in Scotland. The remains of several hundred skeletons have been procured within the last century, from five or six small lakes in Forfarshire, where shell-marl has been worked. Those of the stag (*Cervus elaphus*) are most numerous, and if the others be arranged in the order of their relative abundance, they will follow nearly thus : the ox, the boar, the horse, the sheep, the dog, the hare, the fox, the wolf, and the cat. The beaver seems extremely rare, but it has been found in the shell-marl of Loch Marlie, in Perthshire, and in the parish of Edrom, in Berwickshire.

In the greater part of these lake deposits there are no signs of floods, and the expanse of water was originally so confined, that the smallest of the above-mentioned quadrupeds could have crossed, by swimming, from one shore to the other. Deer, and such species as take readily to the water, may often have been mired in trying to land, where the bottom was soft and quaggy, and, in their efforts to escape, may have plunged deeper into the marly bottom. Some individuals, we suspect,

* Scots Mag., vol. xxxiii.

of different species, have fallen in when crossing the frozen surface in winter, for nothing can be more treacherous than the ice when covered with snow, in consequence of the springs, which are numerous, and which, always retaining an equal temperature, cause the ice, in certain spots, to be extremely thin, while, in every other part of the lake, it is strong enough to bear the heaviest weights.

As the bones of mammalia are often so abundantly preserved in peat, and in such lakes as we have just described, the encroachments of a sea upon a coast may sometimes throw down the imbedded skeletons, so that they may be carried away by tides and currents, and entombed in subaqueous formations. Some of the smaller quadrupeds, also, which burrow in the ground, as well as reptiles and every species of plant, are liable to be cast down into the waves by this cause, which must not be overlooked, although we believe it to be of comparatively small importance amongst the numerous agents whereby terrestrial organic remains may be included in submarine strata.

CHAPTER XVI.

Imbedding of the remains of man and his works in subaqueous strata—Drifting of bodies to the sea by river-inundations—Destruction of bridges and houses— Burial of human bodies in the sea—Loss of lives by shipwreck—Circumstances under which human corpses may be preserved under a great thickness of recent deposits—Number of wrecked vessels—Durable character of many of their contents—Examples of fossil skeletons of men—Of fossil canoes, ships, and works of art—Of the chemical changes which certain metallic instruments have undergone after long submergence—Effects of the subsidence of land in imbedding cities and forests in subaqueous strata—Earthquake of Cutch in 1819— Submarine forests—Berkley's arguments for the recent date of the creation of man—Concluding remarks.

WE shall now proceed to inquire in what manner the mortal remains of man and the works of his hands may be permanently preserved in subaqueous strata. Of the many hundred million human beings which perish in the course of every century on the land, every vestige is usually destroyed in the course of a few thousand years, but of the smaller number that perish in the waters, a considerable proportion must frequently be entombed, under such circumstances, that parts of them may endure throughout entire geological epochs.

We have already seen how the bodies of men, together with those of the inferior animals, are occasionally washed down during river-inundations into seas and lakes, of which we shall now enumerate some additional examples.

Belzoni witnessed a flood on the Nile in September, 1818, where, although the river only rose three feet and a half above its ordinary level, several villages, with some hundreds of men, women, and children, were swept away *. We mentioned in a former volume that a rise of six feet of water in the Ganges in 1763, was attended with a much greater loss of lives.

In the year 1771, at the time of the bursting of the Sol-

* Narrative of Discovery in Egypt, &c. London, 1820.

way moss before alluded to, when the inundations in the
north of England appear to have equalled the recent floods in
Morayshire, a great number of houses and their inhabitants
were swept away by the rivers Tyne, Can, Wear, Tees, and
Greta; and no less than twenty-one bridges were destroyed
in the courses of these rivers. At the village of Bywell the
flood tore the dead bodies and coffins out of the churchyard,
and bore them away, together with many of the living inhabitants. During the same tempest an immense number of
cattle, horses, and sheep, were also transported to the sea,
while the whole coast was covered with the wreck of ships.
Four centuries before (in 1338), the same district had been
visited by a similar continuance of heavy rains followed by
disastrous floods, and it is not improbable that these catastrophes may recur periodically. As the population increases,
and buildings and bridges are multiplied, we must expect that
the loss of lives and property will rather augment *.

If to the hundreds of human bodies committed to the deep
in the way of ordinary burial, we add those of individuals lost
by shipwreck, we shall find that, in the course of a single year,
a great number of human remains are consigned to the subaqueous regions. We shall hereafter advert to a calculation
by which it appears that more than five hundred *British*
vessels alone, averaging each a burden of about one hundred
and twenty tons, are wrecked, and sink to the bottom, *annually*. Of these the crews for the most part escape, although it
sometimes happens that all perish. In one great naval action
several thousand individuals sometimes share a watery grave.

Many of these corpses are instantly devoured by predaceous
fish, sometimes before they reach the bottom ; still more frequently when they rise again to the surface and float in a state
of putrefaction. Many decompose on the floor of the ocean
where no sediment is thrown down upon them, but if they fall
upon a reef where corals and shells are becoming agglutinated
into a solid rock, or subside where the delta of a river is ad-

* Scots Mag., vol. xxxiii. 1771.

vancing, they may be preserved for an incalculable series of ages in these deposits.

Often at the distance of a few hundred feet from a coral reef there are no soundings at the depth of many hundred fathoms. Here if a ship strike and be wrecked, it may soon be covered by calcareous sand and fragments of coral detached by the breakers from the summit of a submarine mountain, and which may roll down to its base. Wrecks are known to have been common for centuries near certain reefs, so that canoes, merchant vessels, and ships of war may have sunk and have been enveloped in these situations in calcareous sand and breccia. Suppose a volcanic eruption to cover such remains with ashes and sand, and that over the tufaceous strata resulting from these ejections, a current of lava is afterwards poured, the ships and human skeletons might then remain uninjured beneath the superincumbent rock, like the houses and works of art in the subterranean cities of Campania. That cases may have already occurred where human remains have been thus preserved in a fossil state beneath masses more than a thousand feet in thickness, is by no means improbable, for in some volcanic archipelagos a period of thirty or forty centuries might well suffice for such an accumulation of matter.

We stated that at the distance of about forty miles from the base of the delta of the Ganges, there is a circular space about fifteen miles in diameter where soundings of a thousand feet sometimes fail to reach the bottom. As during the flood season the quantity of mud and sand poured by the great rivers into the Bay of Bengal, is so great that the sea only recovers its transparency at the distance of sixty miles from the coast, this depression must be gradually shoaling, especially as during the monsoons the sea, loaded with mud and sand, is beaten back in that direction towards the delta. Now if a ship or human body sink down to the bottom in such a spot, it is by no means improbable that it may become buried under a depth of three or four thousand feet of sediment in the same number of years.

Even on that part of the floor of the ocean whither no accession of drift matter is carried, (a part which we believe to constitute, at any given period, by far the larger proportion of the whole submarine area,) there are circumstances accompanying a wreck which favour the conservation of skeletons. For when the vessel fills suddenly with water, especially in the night, many persons are drowned between decks and in their cabins, so that their bodies are prevented from rising again to the surface. The vessel often strikes upon an uneven bottom and is overturned, in which case the ballast consisting of sand, shingle, and rock, or the cargo, frequently composed of heavy and durable materials, may be thrown down upon the carcasses. In the case of ships of war, cannon, shot, and other warlike stores, may press down with their weight the timbers of the vessel when they decay, and beneath these and the metallic substances the bones of man may be preserved.

When we reflect on the number of curious monuments consigned to the bed of the ocean in the course of every naval war from the earliest times, our conceptions are greatly raised respecting the multiplicity of lasting memorials which man is leaving of his labours. During our last great struggle with France, thirty-two of our ships of the line went to the bottom in the space of twenty-two years, besides seven fifty-gun ships, eighty-six frigates, and a multitude of smaller vessels. The navies of the other European powers, France, Holland, Spain, and Denmark, were almost annihilated during the same period, so that the aggregate of their losses must have many times exceeded that of Great Britain. In every one of these ships were batteries of cannon constructed of iron or brass, whereof a great number had the dates and places of their manufacture inscribed upon them in letters cast in metal. In each there were coins of copper, silver, and often many of gold, capable of serving as valuable historical monuments; in each were an infinite variety of instruments of the arts of war and peace, many formed of materials, such as glass and earthenware, capable of lasting for indefinite ages when once removed from the mechanical action of the waves, and buried under a mass of matter which may exclude the corroding action of sea-water.

But the reader must not imagine that the fury of war is more conducive than the peaceful spirit of commercial enter-prise to the accumulation of wrecked vessels in the bed of the sea. From an examination of Lloyd's lists from the year 1793, to the commencement of 1829, it has appeared that the number of *British vessels* alone lost during that period amounted, on an average, to no less than one and a half *daily* *, a greater number than we should have anticipated, although we learn from Moreau's tables that the number of merchant vessels employed at one time in the navigation of Eng-land and Scotland, amounts to about twenty thousand, having one with another a mean burden of one hundred and twenty tons †. Out of five hundred and fifty-one ships of the royal navy lost to the country during the period above mentioned, only one hundred and sixty were taken or destroyed by the enemy, the rest having either stranded or foundered, or having been burnt by accident ‡, a striking proof that the dangers of our naval warfare, however great, may be far exceeded by the storm, the hurricane, the shoal, and all the other perils of the deep.

Millions of dollars and other coins have been sometimes submerged in a single ship, and on these, when they happen to be enveloped in a matrix capable of protecting them from chemical changes, much information of historical interest will remain inscribed and endure for periods as indefinite as have the delicate markings of zoophytes or lapidified plants in some of the ancient secondary rocks. In almost every large ship, moreover, there are some precious stones set in seals, and other articles of use and ornament composed of the hardest sub-stances in nature, on which letters and various images are

* I am indebted to my friend Captain W. H. Smyth, R. N., for this information.
  † Cæsar Moreau's Tables of the Navigation of Great Britain.
  ‡ I give these results on the authority of Captain W. H. Smyth, R. N.

carved—engravings which they may retain when included in
subaqueous strata, as long as a crystal preserves its natural
form.

It was a splendid boast, that the deeds of the English
chivalry at Agincourt made Henry's chronicle

> ———— as rich with praise
> As is the ooze and bottom of the deep
> With sunken wreck and sumless treasuries ;

for it is probable that a greater number of monuments of the
skill and industry of man will, in the course of ages, be collected together in the bed of the ocean, than will be seen at
one time on the surface of the continents.

If our species be of as recent a date as we suppose, it will
be vain to seek for the remains of man and the works of his
hands imbedded in submarine strata, except in those regions
where violent earthquakes are frequent, and the alterations
of relative level so great, that the bed of the sea may have
been converted into land within the historical era. We do not
despair of the discovery of such monuments whenever those
regions which have been peopled by man from the earliest
ages, and which are at the same time the principal theatres of
volcanic action, shall be examined by the joint skill of the
antiquary and the geologist.

There can be no doubt that human remains are as capable
of resisting decay as are the harder parts of the inferior animals; and we have already cited the remark of Cuvier, that
" in ancient fields of battle the bones of men have suffered as
little decomposition as those of horses which were buried in
the same grave *." In the delta of the Ganges bones of men
have been found in digging a well at the depth of ninety
feet † ; but as that river frequently shifts its course and fills
up its ancient channels, we are not called upon to suppose that
these bodies are of extremely high antiquity, or that they were
buried when that part of the surrounding delta where they
occur was first gained from the sea.

* Vol. i. p. 154.   † Hoff., vol. i. p. 379.

Several skeletons of men, more or less mutilated, have been found in the West Indies, on the north-west coast of the main-land of Guadaloupe, in a kind of rock which is known to be forming daily, and which consists of minute fragments of shells and corals, incrusted with a calcareous cement resembling tra-vertin, which has also bound the different grains together. The lens shows that some of the fragments of coral composing this stone, still retain the same red colour which is seen in the reefs of living coral which surround the island. The shells belong to species of the neighbouring sea intermixed with some terrestrial kinds which now live on the island, and among them is the Bulimus Guadaloupensis of Férussac. The human skeletons still retain some of their animal matter, and all their phosphate of lime. One of them, of which the head is wanting, may now be seen in the British Museum, and another in the Royal Cabinet at Paris. According to Mr. König, the rock in which the former is inclosed is harder under the mason's saw and chisel, than statuary marble. It is described as form-ing a kind of glacis, probably an indurated beach, which slants from the steep cliffs of the island to the sea, and is nearly all submerged at high tide.

Similar formations are in progress in the whole of the West Indian archipelago, and they have greatly extended the plain of Cayes in St. Domingo, where fragments of vases and other human works have been found at a depth of twenty feet. In digging wells also near Catania, tools have been discovered in a rock somewhat similar.

When a vessel is stranded in shallow water, it usually be-comes the nucleus of a sand bank, as has been exemplified in several of our harbours, and this circumstance tends greatly to its preservation. About fifty years ago, a vessel from Purbeck, laden with three hundred tons of stone, struck on a shoal off the entrance of Poole harbour and foundered; the crew were saved, but the vessel and cargo remain to this day at the bottom. Since that period the shoal at the entrance of the harbour has so extended itself in a westerly direction towards Peveril Point in Purbeck, that the navigable channel is thrown a mile nearer that Point *. The cause is obvious; the tidal current deposits the sediment with which it is charged around any object which checks its velocity. Matter also drifted along the bottom is arrested by any obstacle, and accumulates round it just as the African sand-winds, before described, raise a small hillock over the carcasses of every dead camel exposed on the surface of the desert.

We alluded, in the former volume †, to an ancient Dutch vessel, discovered in the deserted channel of the river Rother, in Sussex, of which the oak wood was much blackened, but its texture unchanged. The interior was filled with fluviatile silt, as was also the case in regard to a vessel discovered in a former bed of the Mersey, and another disinterred where the St. Catherine Docks are excavated in the alluvial plain of the Thames. In like manner many ships have been found preserved entire in modern strata, formed by the silting up of estuaries along the southern shores of the Baltic, especially in Pomerania. Between Bromberg and Nakel, for example, a vessel and two anchors in a very perfect state were dug up far from the sea ‡.

At the mouth of a river in Nova Scotia, a schooner of thirty-two tons, laden with live stock, was lying with her side to the tide, when the bore, or tidal wave, which rises there about ten feet in perpendicular height, rushed into the estuary and overturned the vessel, so that it instantly disappeared. After the tide had ebbed, the schooner was so totally buried in the sand, that the taffrel or upper rail of the deck was alone visible §. We are informed by Leigh, that, on draining Martin Meer, a lake eighteen miles in circumference, in Lancashire, a bed of marl was laid dry, wherein no fewer than eight canoes were found imbedded. In figure and dimensions they were not unlike those now used in America. In a morass about nine

---

* This account I received from the Honourable A. Harris.

† Vol. i. p. 278.   ‡ Hoff., vol. i. p. 368.

§ Silliman's Geol. Lectures, p. 78, who cites Penn.

miles distant from this Meer, a whetstone and an axe of mixed metal were dug up *. In Ayrshire also, three canoes were found in Loch Doon some few years ago; and during the present year (1831) four others, each hewn out of separate oak trees. They were twenty-three feet in length, two and a half in depth, and nearly four feet in breadth at the stern. In the mud which filled one of them, was found a war club of oak and a stone battle-axe.

The only examples of buried vessels to which we can obtain access, are in such situations as we have mentioned, but we are unable to examine those which have been subjected to great pressure, at the bottom of a deep ocean. It is extremely possible that the submerged wood-work of ships which have sunk where the sea is two or three miles deep, has undergone greater chemical changes in an equal space of time, for the experiments of Scoresby before mentioned show that wood may at certain depths be impregnated in a single hour with salt-water, so that its specific gravity is entirely altered.

It may often happen that hot springs charged with carbonate of lime, silex and other mineral ingredients, may issue at great depths, in which case every pore of the vegetable tissue may be injected with the lapidifying liquid, whether calcareous or siliceous, before the smallest decay commences. The conversion also of wood into lignite is probably more rapid under such enormous pressure. But the change of the timber into lignite or coal would not prevent the original form of a ship from being distinguished, for as we find in strata of the carboniferous era, the bark of the hollow reed-like trees converted into coal, and the central cavity filled with sandstone, so might we trace the outline of a ship in coal, and in the indurated mud, sandstone, or limestone filling the interior, we might discover instruments of human art, ballast consisting of rocks foreign to the rest of the stratum, and other contents of the ship.

Many of the metallic substances which fall into the waters,

* Leigh's Lancashire, p. 17, A. D. 1700.

probably lose, in the course of ages, the forms artificially im-
parted to them ; but under many circumstances these may be
preserved for indefinite periods. The cannon inclosed in a
calcareous rock, drawn up from the delta of the Rhone, which
is now in the museum at Montpellier, might probably have
endured as long as the calcareous matrix ; but even if the metallic matter had been removed and had entered into new combinations, still a mould of its original shape would have been
left, corresponding to those impressions of shells which we see
in rocks, from which all the carbonate of lime has been sub-
tracted.   About the year 1776, says Mr. King, some fishermen sweeping for anchors in the Gull stream, (a part of the
sea near the Downs,) drew up a very curious old swivel gun,
near eight feet in length.   The barrel, which was about five
feet long, was of brass ; but the handle by which it was
traversed, was about three feet in length, and the swivel
and pivot on which it turned were of iron.   Around these
latter were formed incrustations of sand converted into a kind
of stone, of an exceeding strong texture and firmness; whereas
round the barrel of the gun, except where it was near adjoining to the iron, there was no such incrustation, the greater
part of it being clean and in good condition, just as if it had
still continued in use.   In the incrusting stone, adhering to
it on the outside, were a number of shells and corallines,
" just as they are often found in a fossil state."   These
were all so strongly attached, that it required as much force
to separate them from the matrix, " as to break a fragment off
any hard rock *."

In the year 1745, continues the same writer, the Fox man-
of-war was stranded on the coast of East Lothian and went to
pieces.   About thirty-three years afterwards a violent storm
laid bare a part of the wreck, and threw up near the place
several masses " consisting of iron, ropes and balls," covered
over with ochreous sand concreted and hardened into a kind
of stone.   The substance of the rope was very little altered.

* Phil. Trans., 1779.

The consolidated sand retained perfect impressions of parts of
an iron ring, "just in the same manner as impressions of
extraneous fossil bodies are found in various kinds of strata *."

After a storm in the year 1824, which occasioned a considerable shifting of the sands near St. Andrew's, in Scotland, a gun
barrel of ancient construction was found, which is conjectured
to have belonged to one of the wrecked vessels of the Spanish
armada.   It is now in the museum of the Antiquarian Society
of Scotland, and is encrusted over by a thin coating of sand,
the grains of which are cemented by brown ferruginous matter.
Attached to this coating are fragments of various shells, as of
the common cardium, mya, &c.

Many other examples are recorded of iron instruments taken
up from the bed of the sea near the British coasts, incased by a
thick coating of conglomerate, consisting of pebbles and sand,
cemented by oxide of iron.

Dr. Davy describes in the Philosophical Transactions †, a
bronze helmet of the antique Grecian form, taken up in 1825,
from a shallow part of the sea, between the citadel of Corfu
and the village of Castrades.   Both the interior and exterior
of the helmet were partially encrusted with shells, and a deposit
of carbonate of lime.   The surface generally, both under the
incrustation and where freed from it, was of a variegated colour,
mottled with spots of green, dirty white, and red.   On minute
inspection with a lens, the green and red patches proved to
consist of crystals of the red oxide and carbonate of copper,
and the dirty white chiefly of oxide of tin.

The mineralizing process, says Dr. Davy, which has produced these new combinations, has in general penetrated very
little into the substance of the helmet.   The incrustation and
rust removed, the metal is found bright beneath, in some
places considerably corroded, in others very slightly.   It
proves on analysis to be copper alloyed with 18.5 per cent. of
tin.   Its colour is that of our common brass, and it possesses
a considerable degree of flexibility :—

* Phil. Trans., vol. lxix., 1779.   † 1826, part ii. p. 55.

" It is a curious question," he adds, " how the crystals were formed in the helmet, and on the adhering calcareous deposit. There being no reason to suppose deposition from solution, are we not under the necessity of inferring, that the mineralizing process depends on a small motion and separation of the particles of the original compound ? This motion may have been due to the operation of electro-chemical powers which may have separated the different metals of the alloy."

Effects of the Submersion of Land by Earthquakes.

We have hitherto considered the transportation of plants and animals from the land by *aqueous* agents, and their inhumation in lacustrine or submarine deposits, and we may now inquire what tendency the subsidence of tracts of land by *earthquakes* may have to produce analogous effects. Several examples of the sinking down of buildings and portions of towns near the shore to various depths beneath the level of the sea, during subterranean movements, were enumerated in a former volume, when we treated of the changes brought about by *inorganic* causes. The events alluded to were comprised within a brief portion of the historical period, and confined to a small number of the regions of active volcanos. Yet these authentic facts, relating merely to the last century and a half, gave indications of considerable change which must have taken place in the physical geography of the globe. If, during the earthquake of Jamaica in 1692, some of the houses in Port Royal subsided, together with the ground they stood upon, to the depth of twenty-four, thirty-six and forty-eight feet under water, we are not to suppose that this was the only spot throughout the whole range of the coasts of that island or the bed of the surrounding sea which suffered similar depressions. If the quay at Lisbon sank at once to the depth of six hundred feet in 1755, we must not imagine that this was the only point on the shores of the peninsula where similar phenomena might have been witnessed.

If during the short period since South America has been

colonized by Europeans we have proof of alterations of level at the three principal ports on the western shores, Callao*, Valparaiso, and Conception, we cannot for a moment suspect that these cities so distant from each other have been selected as the peculiar points where the desolating power of the earthquake has expended its chief fury. " It would be a knowing arrow that could choose out the brave men from the cowards," retorted the young Spartan, when asked if his comrades who had fallen on the field of battle were braver than he and his fellow prisoners; we might in the same manner remark that a geologist must attribute no small discrimination and malignity to the subterranean force, if he should suppose it to spare habitually a line of coast many thousand miles in length, with the exception of those few spots where populous towns have been erected. If then we consider how small is the area occupied by the sea-ports of this disturbed region,—points where alone each slight change of the relative level of sea and land can be recognized, and reflect on the proofs in our possession of the local revolutions that have happened on the site of each port, within the last century and a half, our conceptions must be greatly exalted respecting the magnitude of the alterations which the Andes may have undergone even in the course of the last six thousand years.

We cannot better illustrate the manner in which a large extent of surface may be submerged, so that the terrestrial plants and animals may become imbedded in subaqueous strata, than by referring to the earthquake of Cutch, in 1819, alluded to by us in a former volume †. We shall enter somewhat more fully into details concerning that catastrophe than the

* It is well known that during the great earthquake of Lima, in 1746, part of the promontory south of Callao sank down, and it is a common story at Lima that its former termination became the present isle of San Lorenzo, between which and the main land there is now a navigable channel. The submerged arches of a church, and the present position of other buildings, are said to indicate that the site of Callao underwent, during the earthquakes, a change of level; an interesting fact, the evidences of which we hope will soon be examined by some of our naval officers, and other intelligent persons frequenting that port.

† Vol. i. p. 405.

immediate subject of the present chapter might require, in order to lay before the reader the information obtained during the recent survey of Cutch.

The published account of Lieutenant A. Burnes *, who examined that portion of the delta of the Indus in 1826 and 1829, confirms the facts before enumerated by us, and furnishes the following important particulars. The tract around Sindree, which subsided during the earthquake in June, 1819, was converted from dry land into sea in the course of a few hours, the new-formed mere extending for a distance of sixteen miles on each side of the fort, and probably exceeding in area the lake of Geneva. Neither the rush of the sea into this new depression, nor the movement of the earthquake, threw down the small fort of Sindree, the interior of which is said to have become *a tank*, the water filling the space within the walls, and the four towers continuing to stand, so that on the day after the earthquake the people in the fort who had ascended to the top of one of the towers saved themselves in boats. Immediately after the shock the inhabitants of Sindree saw, at the distance of five miles from their village, a long elevated mound, where previously there had been a low and perfectly level plain. To this uplifted tract they gave the name of " Ullah bund," or " the Mound of God," to distinguish it from an artificial barrier previously thrown across an arm of the Indus.

It is already ascertained that this newly raised country is *upwards of fifty miles* in length from east to west, running parallel to that line of subsidence before mentioned, which caused the grounds around Sindree to be flooded. The range of this elevation extends from Puchum island towards Gharee; its breadth from north to south is conjectured to be in some parts *sixteen miles*, and its greatest ascertained height above the original level of the delta is ten feet, an elevation which appears to the eye to be very uniform throughout.

For several years after the convulsion of 1819, the course of the Indus was very unsettled, and at length in 1826, the river burst its banks above Sinde, and forcing its way in a more

* Now in the Library of the Royal Asiatic Society.

direct course to the sea, cut right through the " Ullah bund," whereby a natural section was obtained. In the perpendicular cliffs thus laid open, Lieutenant Burnes found that the upraised land consisted of beds of clay filled with shells. The new channel of the river, where it intersected the " bund," was eighteen feet deep, and during the swells in 1826, it was two or three hundred yards in width, but in 1828 the channel was still further enlarged. The Indus, when it first opened this new passage, threw such a body of water into the new lake or salt lagoon of Sindree, that it became fresh for many months, but it had recovered its saltness in 1828, when the supply of river-water was less copious, and finally it became more salt than the sea, in consequence, as the natives suggested to Lieutenant Burnes, of the saline particles with which " the Runn of Cutch " is impregnated.

Besides *Ullah bund*, there appears to have been another elevation south of Sindree, parallel to that before mentioned, respecting which, however, no exact information has yet been communicated. There is a tradition of an earthquake, which, about three centuries before, upheaved a large area of the bed of the sea, and converted it into land in the district now called " the Runn," so that numerous harbours were laid dry and ships were wrecked and engulphed ; in confirmation of which account it was observed in 1819, that in the jets of black muddy water thrown out of fissures in that region, there were cast up numerous pieces of wrought iron and ship nails.

We must not conclude without alluding to a *moral* phenomenon connected with this tremendous catastrophe, which we regard as highly deserving the attention of geologists. The author above cited states that " these wonderful events passed *unheeded* by the inhabitants of Cutch," for the region convulsed, though once fertile, had for a long period been reduced to sterility by want of irrigation, so that the natives were indifferent as to its fate. Now it is to this profound apathy, which all but highly civilized nations feel in regard to physical events, not having an immediate influence on their worldly fortunes, that we must ascribe the extraordinary dearth of historical information concerning changes of the earth's surface, which modern observations show to be by no means of rare occurrence in the ordinary course of nature.

It is stated that, for some years after the earthquake, the withered tamarisks and other shrubs protruded their tops above the waves, in parts of the submerged tract around Sindree; but after the flood of 1826 they were seen no longer. Every geologist will at once perceive that forests sunk by such subterranean movements, may become imbedded in subaqueous deposits both fluvatile and marine, and the trees may still remain erect, or sometimes the roots and part of the trunks may continue in their original position, while the current may have broken off, or levelled with the ground, their upper stems and branches.

But although a certain class of geological phenomena may be referred to the repetition of such catastrophes, we must hesitate before we call in to our aid the action of earthquakes, to explain what have been termed submarine forests, observed at various points around the shores of Great Britain. We have already hinted that the explanation of some of these may be sought in the encroachments of the sea, in estuaries, and the varying level of the tides, at distant periods on the same parts of our coast*. After examining, in 1829, the so called submarine forest of Happisborough in Norfolk, I found that it was nothing more than a tertiary lignite of the " Crag " period, which becomes exposed in the bed of the sea as soon as the waves sweep away the superincumbent strata of bluish clay. So great has been the advance of the sea upon our eastern shores within the last eight centuries, that whenever we find a mass of submerged timber near the sea side, or at the foot of the existing cliffs which we cannot suppose to be a mere accumulation of drift, vegetable matter, we should endeavour to find a solution of the problem, by reference to any cause rather than an earthquake. For we can scarcely doubt that the present outline of our coast, the shape of its estuaries, and the formation of its cliffs are of very modern date, probably within the human era, whereas we have no reason whatever to imagine that this part of Europe has been agitated by subterranean convulsions, capable of altering the relative level of land and sea, at so extremely recent a period.

Some of the buildings which have at different times subsided beneath the level of the sea, have been immediately covered up to a certain extent with strata of volcanic matter showered down upon them. Such was the case at Tomboro in Sumbawa, in the present century, and at the site of the Temple of Serapis, in the environs of Puzzuoli, probably in the 12th century. The entrance of a river charged with sediment in the vicinity, may still more frequently occasion the rapid envelopement of buildings in regularly stratified formations. But if no foreign matter be introduced, the buildings when once removed to a depth where the action of the waves is insensible, and where no great current happens to flow, may last for indefinite periods, and be as durable as the floor of the ocean itself, which may often be composed of the very same materials. There is no reason to doubt the tradition mentioned by the classic writers, that the submerged Grecian towns of Bura and Helice were seen under water; and I am informed by an eye-witness that eighty-eight years after the convulsion of 1692, the houses of Port Royal were still visible at the bottom of the sea *.

* Admiral Sir Charles Hamilton frequently saw the submerged houses of Port Royal in the year 1780, in that part of the harbour which lies between the town and the usual anchorage of men-of-war. Bryan Edwards also says in his History of the West Indies, (vol. i. p. 235, oct. ed. 3 vols., 1801,) that in 1793 the *ruins* were visible in clear weather from the boats which sailed over them. I regret to see that Mr. De la Beche, in his valuable Manual of Geology, (p. 130,) has evinced so much scepticism in regard to the accuracy of the evidence collected by Sir Hans Sloane, respecting the catastrophe of Port Royal, a town with which Sir H. was well acquainted. To me the original documents collected immediately after the event, appear to bear the intrinsic stamp of truth. The objection against the fact alleged by several eye-witnesses, " that the chimney tops alone of many houses were seen after the shocks, as well as the masts of vessels just projecting above the waves," is quite futile. Perhaps the chimneys in Port Royal might in 1692, have been confined to low kitchens, as Mr. De la Beche says they now are, and they might only have been fifteen or twenty feet in height, still the same subsidence which reduced them to the level of the water might cause the

We cannot conclude this chapter without recalling to the reader's mind a memorable passage written by Berkely a century ago, in which he inferred, on grounds which may be termed strictly geological, the recent date of the creation of man. " To any one," says he, " who considers that on digging into the earth such quantities of shells, and in some places bones and horns of animals are found sound and entire, after having lain there in all probability some thousands of years ; it should seem probable that guns, medals and implements in metal or stone might have lasted entire, buried under ground forty or fifty thousand years if the world had been so old. How comes it then to pass that no remains are found, no antiquities of those numerous ages preceding the Scripture accounts of time ; that no fragments of buildings, no public monuments, no intaglias, cameos, statues, basso-relievos, medals, inscriptions, utensils, or artificial works of any kind are ever discovered, which may bear testimony to the existence of those mighty empires, those successions of monarchs, heroes, and demi-gods for so many thousand years ? Let us look forward and suppose ten or twenty thousand years to come, during which time we will suppose that plagues, famine, wars and *earthquakes* shall have made great havoc in the world, is it not highly probable that at the end of such a period, pillars, vases, and statues now in being of granite, or porphyry, or jasper, (stones of such hardness as we know them to have lasted two thousand years above ground, without any considerable alteration) would bear record of these and past ages ? Or that some of our current coins might then be dug up, or old walls and the foundations of buildings shew themselves, as well as the shells and stones of *the primeval world*, which are preserved down to our times * ? "

ships *which were previously floating* to disappear entirely, with the exception of the tops of their masts. Besides, we infer from the various narratives, that the subsidences were very unequal at different neighbouring points.

I have great pleasure in stating, that on my requesting Mr. De la Beche to send me more exact particulars, respecting the present state of the harbour of Port Royal, he has ordered a survey to be made.

* Alciphron, or the Minute Philosopher, vol. ii. pp. 84, 85. 1732.

That many signs of the agency of man would have lasted at least as long as " the shells of the primeval world," had our race been so ancient, we are as fully persuaded as Berkely ; and we anticipate with confidence that many edifices and implements of human workmanship, and the skeletons of men, and casts of the human form, will continue to exist when a great part of the present mountains, continents, and seas have disappeared. Assuming the future duration of the planet to be indefinitely protracted, we can foresee no limit to the perpetuation of some of the memorials of man, which are continually entombed in the bowels of the earth or in the bed of the ocean, unless we carry forward our views to a period sufficient to allow the various causes of change both igneous and aqueous, to remodel more than once the entire crust of the earth. *One* complete revolution will be inadequate to efface every monument of our existence, for many works of art might enter again and again into the formation of successive eras, and escape obliteration even though the very rocks in which they had been for ages imbedded were destroyed, just as pebbles included in the conglomerates of one epoch often contain the organized remains of beings which flourished during a prior era.

Yet it is no less true, as a late distinguished philosopher has declared, " that none of the works of a mortal being can be eternal *." They are in the first place wrested from the hands of man, and lost as far as regards their subserviency to his use, by the instrumentality of those very causes which place them in situations where they are enabled to endure for indefinite periods. And even when they have been included in rocky strata, when they have been made to enter as it were into the solid framework of the globe itself, they must nevertheless eventually perish, for every year some portion of the earth's crust is shattered by earthquakes or melted by volcanic fire, or ground to dust by the moving waters on the surface. " The river of Lethe," as Bacon eloquently remarks, " runneth as well above ground as below †."

* Davy, Consolations in Travel, p. 276.
† Essay on the Vicissitude of Things.

WE have hitherto treated of the imbedding of terrestrial plants and animals, and of human remains in the deposits that are now forming beneath the waters, and we come next to consider in what manner *aquatic* species may be entombed in strata, formed in their own element.

Imbedding of Fresh-water Plants and Animals.

The remains of species belonging to those genera of the animal and vegetable kingdoms, which are more or less exclusively confined to fresh-water, are for the most part preserved in the beds of lakes or estuaries, but they are oftentimes swept down by rivers into the sea, and there intermingled with the exuviæ of marine races. The phenomena attending their inhumation in lacustrine deposits, may be sometimes revealed to our observation by the drainage of small lakes, such as are those in Scotland which have been laid dry for the sake of obtaining shell marl for agricultural uses.

In these recent formations, as seen in Forfarshire, two or three beds of calcareous marl are sometimes observed separated from each other by layers of drift peat, sand, or fissile clay. The marl often consists almost entirely of an aggregate of shells of the genera limnea, planorbis, valvata, and cyclas, with some few others, species of all which now exist in Scotland. A considerable proportion of the testacea appear to have died very young, and few of the shells are of a size which indicates their having attained a state of maturity. The shells are sometimes entirely decomposed, forming a pulverulent marl; sometimes they are in a state of good preservation. They are frequently intermixed with stems of charæ and other aquatic vegetables, which are matted together and compressed, forming laminæ often as thin as paper.

As the chara is an aquatic plant, which occurs frequently fossil in formations of different eras, and is often of much importance to the geologist in characterizing entire groups of strata, we shall describe the manner in which the recent species have been found in a petrified state. They occur in one of the lakes of Forfarshire, inclosed in nodules, and sometimes in a continuous stratum of a kind of travertin.

The seed-vessel of these plants is remarkably tough and hard, and consists of a membranous nut covered by an integument (fig. *d* diagram No. 2,) both of which are spirally striated or ribbed. The integument is composed of five spiral valves,

(No. 2.)

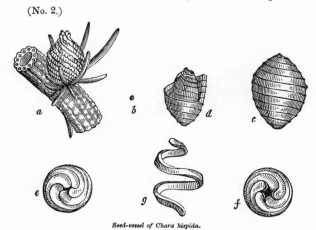

Seed-vessel of Chara hispida.

(*a*) Part of the stem with the seed-vessel attached. Magnified.
(*b*) Natural size of the seed-vessel.
(*c*) Integument of the Gyrogonite, or petrified seed-vessel of Chara hispida, found in the Scotch marl-lakes. Magnified.
(*d*) Section showing the nut within the integument.
(*e*) Lower end of the integument to which the stem was attached.

of a quadrangular form (fig. *g*). In *Chara hispida* which abounds in the lakes of Forfarshire, and which has become fossil in the Bakie Loch, each of the spiral valves of the seed-vessel turns rather more than twice round the circumference, the whole together making between ten and eleven rings. The number of these rings differs greatly in different species, but in the same appears to be very constant.

The stems of charæ occur fossil in the Scotch marl in great abundance. In some species, as in *Chara hispida*, the plant when living contains so much carbonate of lime in its vegetable organization, independently of calcareous incrustation, that it effervesces strongly with acids when dried. The stems of *Chara hispida* are longitudinally striated, with a tendency to be spiral. These striæ, as appears to be the case with all charæ, turn always like the worm of a screw from right to left, while those of the seed-vessel wind round in a contrary direction. A cross section of the stem exhibits a curious structure, for it is composed of a large tube surrounded by smaller tubes, (diagram No. 3. fig. *b, c,*) as is seen in some extinct, as well

(No. 3.)

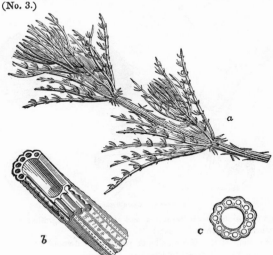

Stem and branches of Chara hispida.

(*a*)   Stem and branches of the natural size.
(*b*)   Section of the stem magnified.
(*c*)   Showing the central tube surrounded by two rings of smaller tubes.

as recent species.   In the stems of several species, however, there is only a single tube*.

The valves of a small animal called cypris (*C. ornata Lam.*) occur completely fossilized like the stems of charæ, in the Scotch travertin above mentioned.   This cypris inhabits the lakes and ponds of England where it is not uncommon. Species of the same genus also occur abundantly in ancient fresh-water formations.

The recent strata of lacustrine origin above alluded to are of very small extent, but analogous deposits on the grandest scale have been formed in the great lakes of North America.   By the subsidence of the waters of Lakes Superior and Huron, occasioned probably by the partial destruction of their barriers at some unknown period, beds of sand one hundred and fifty feet thick are exposed, below which are seen beds of clay, inclosing shells of the very species which now inhabit the lake †.

But no careful examination appears as yet to have been made of recent fresh-water formations within the tropics, where the waters teem with life, and where in the bed of a newly drained lake the remains of the alligator, crocodile tortoise, and perhaps some large fish might be discovered.

Imbedding of Fresh-water Species in Estuary and Marine Deposits.

We have sometimes an opportunity of examining the de posits which within the historical period have silted up some of our estuaries; and excavations made for wells and other purposes, where the sea has been finally excluded, enable us to observe the state of the organic remains in these tracts. The valley of the Ouse between Newhaven and Lewes is one of several estuaries from which the sea has retired within the last seven or eight centuries; and here it appears from the researches of Mr. Mantell, that strata of thirty feet and upwards in thickness have accumulated.   At the top, beneath

---

* Geol. Trans., vol. ii., second series, p. 73. On Fresh-water Marl, &c. By C. Lyell, Esq.   † Dr. Bigsby's Journal of Science, &c. No. 37, pp. 262, 263.

the vegetable soil, is a bed of peat about five feet thick, enclosing many trunks of trees. Next below is a stratum of blue clay containing fresh-water shells of about nine species, such as now inhabit the district. Intermixed with these was observed the skeleton of a deer. Lower down, the layers of blue clay contain with the above-mentioned fresh-water shells several marine species well known on our coast. In the lowest beds, often at the depth of thirty-six feet, these marine testacea occur without the slightest intermixture of fluviatile species, and amongst them the skull of the narwal, or sea-unicorn (*Monodon monoceros*), has been detected. Underneath all these deposits is a bed of pipe-clay, derived from the subjacent chalk *.

If we had no historical information respecting the former existence of an inlet of the sea in this valley, and of its gradual obliteration, the inspection of the section above described would show, as clearly as a written chronicle, the following sequence of events. First, there was a salt-water estuary peopled for many years by species of marine testacea identical with those now living, and into which some of the larger cetacea occasionally entered. Secondly, the inlet grew shallower, and the water became brackish, or alternately salt and fresh, so that the remains of fresh-water and marine shells were mingled in the blue argillaceous sediment of its bottom. Thirdly, the shoaling continued until the river water prevailed, so that it was no longer habitable by marine testacea, but fitted only for the abode of fluviatile species and aquatic insects. Fourthly, a peaty swamp or morass was formed where some trees grew, or, perhaps, were drifted during floods, and where terrestrial quadrupeds were mired. Finally, the soil being only flooded by the river at distant intervals, became a verdant meadow.

We have stated when speaking of the delta of the Ganges, that on the sea-coast there are eight great openings, each of

* Mantell, Geol. of Sussex, p. 285 ; also Catalogue of Org. Rem., Geol. Trans., v. iii., part 1, p. 201. Second Series.

which has evidently, at some ancient periods, served in its turn as the principal channel of discharge. Now as the base of the delta is two hundred miles in length, it must happen that as often as the great volume of river-water is thrown in by a new mouth, the waters of the sea will at one point be converted from salt to fresh, and at another from fresh to salt; for, with the exception of those parts where the principal discharge takes place, the salt-water not only washes the base of the delta, but enters far into every creek and lagoon. It is evident then that repeated alternations of beds containing fresh-water shells, with others filled with corals and marine exuviæ, may here be formed, and each series may be of great thickness, as the sea on which the Gangetic delta gains, is of considerable depth, and intervals of centuries elapse between each alteration in the course of the principal stream.

It is evident that analogous phenomena must sometimes be occasioned by such alternate elevation and depression of the land as was shown in the last chapter to be taking place in the delta of the Indus. But the subterranean movements affect but a small number of the deltas formed at one period on the globe; whereas, the silting up of some of the arms of great rivers and the opening of others, and the consequent variation of the points at which the chief volume of their waters enters the sea, are phenomena common to almost every delta.

The variety of species of testacea contained in the recent calcareous marl of Scotland, before mentioned, is very small, but the abundance of individuals is extremely great, a circumstance which characterizes fresh-water formations in general as compared to marine; for in the latter, as is seen on sea-beaches, coral reefs, or in the bottom of seas examined by dredging, wherever the individual shells are exceedingly numerous there rarely fails to be a vast variety of species.

Imbedding of the Remains of Marine Plants and Animals.

Marine Plants.—We have alluded to the large banks of drift sea-weed which occur on each side of the equator in the

Atlantic, Pacific, and Indian oceans *. These when they subside may often produce considerable beds of vegetable matter.
In Holland submarine peat is derived from fuci, and on parts
of our own coast from *Zostera marina*. In places where
algæ do not generate peat, they may nevertheless leave traces
of their form imprinted on argillaceous and calcareous mud, as
they are usually very tough in their texture.

Cetacea.—It is not uncommon for the larger cetacea, which
can only float in a considerable depth of water, to be carried
during storms or high tides into estuaries, or upon low shores,
where, upon the retiring of high water, they are stranded.
Thus a narwal (*Monodon monoceros*) was found on the beach
near Boston, in Lincolnshire, in the year 1800, the whole of
its body buried in the mud. A fisherman going to his boat
saw the horn and tried to pull it out, when the animal began
to stir itself †. An individual of the common whale (*Balæna
mysticetus*), which measured seventy feet, came ashore near
Peterhead, in 1682. Many individuals of the genus Balæ-
noptera have met the same fate. We may content ourselves
with referring to those cast on shore near Burnt Island, and
at Alloa, recorded by Sibbald and Neill. The other individual mentioned by Sibbald, as having come ashore at Boyne,
in Banffshire, was probably a Razor-back. Of the genus Ca-
todon (*Cachalot*), Ray mentions a large one stranded on the
west coast of Holland in 1598, and the fact is also commemo-
rated in a Dutch engraving of the time of much merit. Sibbald, too, records that a herd of Cachalots, upwards of one
hundred in number, were found stranded at Kairston, Orkney ‡.
The dead bodies of the larger cetacea are sometimes found
floating on the surface of the waters, as was the case with the
immense whale exhibited in London in 1831. And the carcass
of a sea-cow or Lamantine (*Halicora*) was, in 1785, cast ashore
near Leith. We might enumerate many more examples de-

* Page 78.

† Fleming's Brit. Animals, p. 37; in which work may be seen many other
cases enumerated.

rived from foreign as well as British shores, but the facts above cited will suffice to show that such occurrences are not rare.

To some accidents of this kind, we may refer the position of the skeleton of a whale seventy-three feet long, which was found at Airthrey, on the Forth near Alloa, imbedded in clay twenty feet higher than the surface of the highest tide of the river Forth at the present day. From the situation of the Roman station and causeways at a small distance from the spot, it is concluded that the whale must have been stranded there at a period prior to the Christian era *.

Other fossil remains of this class have also been found in estuaries, known to have been silted up in recent times, one example of which we have already mentioned near Lewes, in Sussex. When we reflect on the facility with which these marine mammalia are thus shown to run aground upon shoals, even when there have been no great convulsions, such as hurricanes or earthquakes extending under the ocean, but merely such disturbances as the tides and storms of our seas may cause, we may be better enabled to form a sound opinion, in regard to the probability of certain geological theories, which have acquired no small share of popularity. It has been suggested, that if the ocean, displaced by the sudden upheaving of some great mountain-chain, such as the Andes, should make a transient passage over the land, a covering of alluvium might be left strewed over the hills and valleys, and that, in this alluvium, might be contained the remains of mammalia exclusively terrestrial. The skeleton of the gigantic whale, the long horn of the narwal (harder than ivory), the strong grinders of the lamantine, these and other marine relics of the era

> Omne cum Proteus pecus egit altos
> Visere montes,

might, we are told, be entirely wanting. Not one of them would be conspicuous amongst the refuse of the " bated and retiring flood," but instead of them we should discover the bones, tusks, and teeth of the elephant or rhinoceros, the hip-

* Quart. Journ. of Lit. Sci., &c.   No. 15, p. 172.   Oct. 1819.

popotamus, ox, and horse, with occasionally, perhaps, some intermixture of terrestrial and lacustrine shells! Such, we are taught, would be the memorials of a marine deluge sweeping over our continents! We are, however, willing to admit that they who invent causes without reference to known analogies, are guilty of no inconsistency when they claim some license in the use which they make of their extraordinary agents. If we allow them to " call spirits from the vasty deep " to do their bidding, and to uplift colossal chains, like the Andes, suddenly within the historical era, we must not complain that the effects of such mighty powers are not always such as the analogy of the ordinary laws of Nature would have led us to anticipate.

Marine Testacea.—The aquatic animals and plants which inhabit an estuary are liable, like the trees and land animals which people the alluvial plains of a great river, to be swept from time to time far into the deep. For as a river is perpetually shifting its course, and undermining a portion of its banks with the forests which cover them, so the marine current alters its direction from time to time, and bears away the banks of sand and mud, against which it turns its force. These banks may consist in great measure of shells peculiar to shallow, and sometimes brackish water, which may have been accumulating for centuries, until at length they are carried away and spread out along the bottom of the sea, at a depth at which they could not have lived and multiplied. Thus littoral and estuary shells are more frequently liable even than freshwater species, to be intermixed with the exuviæ of pelagic tribes.

After the late storm of February 4, 1831, when several vessels were wrecked in the estuary of the Forth, the current was directed against a bed of oysters with such force, that great heaps of them were thrown *alive* upon the beach, and remained above high-water mark. Many of these oysters, as also the common whelks (*buccina*), which were thrown up with them, in a living state, were worn by the long attrition of sand which had passed over them as they lay in their native bed, and which had evidently not resulted from the mere action of the tempest by which they had been cast ashore.

From these facts we may learn that the union of the two parts of a bivalve shell does not prove that it may not have been transported to a certain distance; and when we find shells worn, and with all their prominent parts rubbed off, they may still have been imbedded where they grew.

It sometimes appears extraordinary when we observe the violence of the breakers on our coast, and see the strength of the current in removing cliffs and sweeping out new channels, that many tender and fragile shells should inhabit the sea in the immediate vicinity of this turmoil. But a great number of the bivalve testacea, and many also of the turbinated univalves burrow in sand or mud. The solen and the cardium, for example, which are usually found in shallow water near the shore, pierce through a soft bottom without injury to their shells; and the pholas can drill a cavity through mud of considerable hardness. The species of these and many other tribes can sink, when alarmed, with considerable rapidity, often to the depth of several feet, and can also penetrate upwards again to the surface if a mass of matter be heaped upon them. The hurricane, therefore, may expend its fury in vain, and may sweep away even the upper part of banks of sand or mud, or may roll pebbles over them, and yet these testacea may remain below secure and uninjured.

We have already stated that at the depth of nine hundred and fifty fathoms between Gibraltar and Ceuta, Captain Smyth found a gravelly bottom, with fragments of broken shells carried thither probably from the comparatively shallow parts of the neighbouring straits, through which a powerful current flows. Beds of shelly sand might here, in the course of ages, be accumulated several thousand feet thick. But, without the aid of the drifting power of a current, shells may accumulate in the spot where they live and die, at great depths from the surface, if sediment be thrown down upon them; for, even in our own colder latitudes, the depths at which living marine animals abound is very considerable. Captain Vidal ascertained, by soundings lately made off Tory island, on the northwest coast of Ireland, that crustacea, star-fish, and testacea, occurred at various depths between fifty and one hundred fathoms; and in the tropics testacea and zoophytes have been found still deeper.

During the survey of the west coast of Africa, now in progress, Captain Belcher found, by frequent soundings between the twenty-third and twentieth degrees of north latitude, that the bottom of the sea at the depth of from twenty to about fifty fathoms, consists of sand, with a great intermixture of shells often entire, but sometimes finely comminuted. Between the eleventh and ninth degrees of north latitude, on the same coast, at soundings varying from twenty to about eighty fathoms, he brought up abundance of corals and shells mixed with sand. These also were in some parts entire, and in others worn and broken.

In all these cases it is only necessary that there should be some deposition of sedimentary matter, however minute, such as may be supplied by rivers or currents preying on a line of cliffs, and stratified formations, hundreds of feet in thickness, will result in the course of ages, containing throughout organic remains, in a more or less perfect state of preservation.

Corals and Coral reefs.

THE powers of the organic creation in modifying the form and structure of those parts of the earth's crust, which may be said to be undergoing repair, or where new rock-formations are continually in progress, are most conspicuously displayed in the labours of the coral animals. We may compare the operation of these zoophytes in the ocean, to the effects produced on a smaller scale upon the land, by the plants which generate peat. In the case of the Sphagnum, the upper part vegetates while the lower portion is entering into a mineral mass, where the traces of organization usually remain, but in which life has entirely ceased. In the corals, in like manner, the more durable materials of the generation that has passed away, serve as the foundation on which living animals are continuing to rear a similar structure.

The calcareous masses usually termed coral reefs, are by no means exclusively composed of zoophytes, but also a great variety of shells; some of the largest and heaviest of known species contributing to augment the mass. In the south Pacific, great beds of oysters, mussels, *pinnæ marinæ*, and other shells, cover in great profusion almost every reef; and, on the beach of coral islands, are seen the shells of echini and the broken fragments of crustaceous animals. Large shoals of fish also are discernible through the clear blue water, and their teeth and hard palates are probably preserved, although a great portion of their soft cartilaginous bones may decay.

Of the numerous species of zoophytes which are engaged in the production of coral banks, some of the most common belong to the genera Meandrina, Caryophyllia and Astrea, but especially the latter.

The reefs, which just raise themselves above the level of the sea, are usually of a circular or oval form, and are surrounded by a deep and often unfathomable ocean. In the centre of each, there is usually a comparatively shallow lagoon where there is still water, and where the smaller and more delicate kinds of zoophytes find a tranquil abode, while the more strong species live on the exterior margin of the isle, where a great surf usually breaks. When the reef, says M. Chamisso, a naturalist who accompanied Kotzebue, is of such a height that it remains almost dry at low water, the corals leave off building. A continuous mass of solid stone is seen composed of the shells of molluscs and echini, with their broken off prickles and fragments of coral, united by the burning sun, through the medium of the cementing calcareous sand, which has arisen from the pulverization of shells. Fragments of coral limestone are thrown up by the waves, until the ridge becomes so high, that it is covered only during some seasons of the year by the high tides. The heat of the sun often penetrates the mass of stone when it is dry, so that it splits in many places. The force of the waves is thereby enabled to separate and lift blocks of coral, frequently six feet long and three or four in thickness, and throw them upon the reef. " After this the calcareous sand lies undisturbed, and offers to the seeds of trees and plants cast upon it by the waves, a soil upon which they rapidly grow, to overshadow its dazzling white surface. Entire trunks of trees, which are carried by the rivers from other countries and islands, find here, at length, a resting-place after their long wanderings: with these come some small animals, such as lizards and insects, as the first inhabitants. Even before the trees form a wood, the sea-birds nestle here; strayed land-birds take refuge in the bushes; and, at a much later period, when the work has been long since completed, man also appears, builds his hut on the fruitful soil formed by the corruption of the leaves of the trees, and calls himself lord and proprietor of this new creation *."

The Pacific ocean throughout, a space comprehended between the thirtieth parallel of latitude on each side of the equator, is extremely productive of coral. The Arabian gulf is rapidly filling with the same, and it is said to abound in the Persian gulf. Between the coast of Malabar and that of Madagascar, there is also a great sea of coral. Flinders describes an unbroken reef three hundred and fifty miles in length, on the east coast of New Holland; and, between that country and New Guinea, Captain P. King found the coral formations to extend throughout a distance of seven hundred miles, interrupted by no intervals exceeding thirty miles in length.

The chain of coral reefs and islets, called the Maldivas, situated in the Indian ocean to the south-west of Malabar, form a chain four hundred and eighty geographical miles in length, running due north and south. It is composed throughout of a series of circular assemblages of islets, the larger groups being from forty to fifty miles in their longest diameter. Captain Horsburgh, whose chart of these islands is subjoined, informs me that outside of each circle or atoll, as it is termed, there are coral reefs sometimes extending to the distance of two or three miles, beyond which there are no soundings at immense depths. But in the centre of each atoll there is a lagoon from fifteen to twenty fathoms deep. In the channels between the atolls, no soundings have been obtained at the depth of one hundred and fifty fathoms.

The Laccadive islands run in the same line with the Mal-

* Kotzebue's Voyages, 1815-18, vol. iii. p. 331—3.

No. 4.

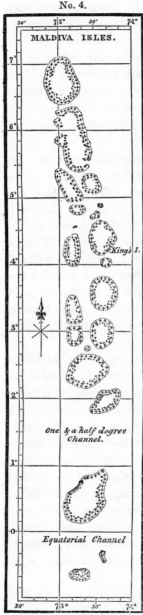

divas, on the north, as do the isles of the Chagos Archipelago, on the south, so that these may be continuations of the same chain of submarine mountains, crested in a similar manner by coral limestone. It would be rash to hazard the hypothesis, that they are all the summits of volcanos, yet we might imagine, that if Java and Sumatra were submerged, they would give rise to a somewhat similar shape in the bottom of the sea; for the volcanos of those islands observe a linear direction, and are often separated from each other by intervals, corresponding to the atolls of the Maldivas; and as they rise to various heights, from five to ten thousand feet above their base, they might leave an unfathomable ocean in the intermediate spaces.

In regard to the thickness of the masses of coral, MM. Quoy and Gaimard are of opinion, that the species which contribute most actively to the formation of solid masses do not grow where the water is deeper than twenty-five or thirty feet. But the branched madrepores, which live at considerable depth, may form the first foundation of a reef, and raise a platform on which other species may build[*], and the sand and

* Journ. of Geograph. Soc. of London, 1831, p. 218.

broken fragments washed by the waves from reefs may, in time, produce calcareous rocks of great thickness.

The rapidity of the growth of coral is by no means great, according to the report of the natives to Captain Beechey. In an island west of Gambier's group, our navigators observed the *Chama gigas* (Tridacna, Lam.) while the animal was yet living, so completely overgrown by coral, that a space only of two inches was left for the extremity of the shell to open and shut *. But conchologists suppose, that the chama may require thirty years or more to attain its full size, so that the fact is quite consistent with a very slow rate of increase in the calcareous reefs. In the late expedition to the Pacific no positive information could be obtained, of any channel having been filled up within a given period, and it seems established that several reefs had remained for more than half a century, at about the same depth from the surface.

The increase of coral limestone, however, may vary greatly according to the sites of mineral springs, for these we know often issue in great numbers at the bottom of the sea in volcanic regions, as in the Mediterranean, for example, where they sometimes cause the sea at great depths to be fresher than at the surface, a phenomenon also declared by the South Sea islanders to be common in the Pacific.

But when we admit the increase of coral limestone to be slow, we are merely speaking with relation to the periods of human observation. It often happens, that parasitic testacea live and die on the shells of the larger slow-moving gasteropods in the South Seas, and become entirely inclosed in an incrustation of compact limestone, while the animal, to whose habitation they are attached, crawls about and bears upon his back these shells, which may be considered as already fossilized. It is, therefore, probable, that the reefs increase as fast as is compatible with the thriving state of the organic beings which chiefly contribute to their formation; and if the rate of augmentation thus implied be called, in conformity to our ordinary ideas of time, gradual and slow, it does not diminish, in the

---

* Beechey's Voyage to the Pacific, &c. p. 157.

least degree, the geological importance of such calcareous masses.

Suppose the ordinary growth of coral limestone to amount to six inches in a century, it will then require three thousand years to produce a reef fifteen feet thick ; but have we any ground for presuming that, at the end of that period, or of ten times thirty centuries, there will be a failure in the supply of lime, or that the polyps and molluscs will cease to act, or that the hour of the dissolution of our planet will first arrive, as the earlier geologists were fain to anticipate ?

Instead of contemplating the brief annals of human events, let us turn to some natural chronometers, to the volcanic isles of the Pacific, for example, which shoot up ten or fifteen thousand feet above the level of the ocean. These islands bear evident marks of having been produced by successive volcanic eruptions ; and coral reefs are sometimes found on the volcanic soil, reaching for some distance from the sea-shore into the interior. When we consider the time required for the accumulation of such mountain masses of igneous matter according to the analogy of known volcanic agency, all idea of extenuating the comparative magnitude of coral limestones, on the ground of the slowness of the operations of lithogenous polyps, must instantly vanish.

The information collected during the late expedition to the Pacific throws much additional light on the peculiarities of form and structure of coral islands. Of thirty-two of these, examined by Captain Beechey, the largest was thirty miles in diameter, and the smallest less than a mile. They were of various shapes, all formed of living coral, except one, which, although of coral formation, was raised about eighty feet above the level of the sea, and encompassed by a reef of living coral. All were increasing their dimensions by the active operations of the lithophytes which appeared to be gradually extending and bringing the immersed parts of their structure to the surface. Twenty-nine of the number had lagoons in their centres, which had probably existed in the others, until they were filled, in the course of time, by zoophytic and other substances.

In the above-mentioned islands, the strips of dry coral encircling the lagoons when divested of loose sandy materials heaped upon them, are rarely elevated more than two feet above the level of the sea; and were it not for the abrupt descent of the external margin which causes the sea to break upon it, these strips would be wholly inundated. " Those parts of the strip which are beyond the reach of the waves are no longer inhabited by the animals that reared them, but have their cells filled with a hard calcareous substance, and present a brown rugged appearance. The parts which are still immersed, or are dry at low water only, are intersected by small channels, and are so full of hollows that the tide, as it recedes, leaves small lakes of water upon them. The width of the plain or strip of dead coral, in the islands which fell under our observation, in no instance exceeded half a mile from the usual wash of the sea to the edge of the lagoon, and in general was only about three or four hundred yards *." Beyond these limits the sides of the island descend rapidly, apparently by a succession of inclined ledges, each terminating in a precipice. The depth of the lagoons is various; in some entered by Captain Beechey, it was from twenty to thirty-eight fathoms.

In the annexed cut (No. 5), one of these circular islands is

No. 5.

View of Whitsunday Island †.

* Captain Beechey, part i. p. 188.
† This plate and the section which follows are copied, by permission of Captain Beechey, from the illustrations of his valuable work before alluded to.

represented just rising above the waves, covered with the cocoa-nut and other trees, and inclosing within, a lagoon of tranquil water.

The accompanying section will enable the reader to comprehend the usual form of such islands.   (No. 6.)

No. 6.

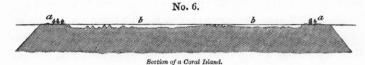

Section of a Coral Island.

(*a a*)   Habitable part of the island, consisting of a strip of coral, inclosing the lagoon.   (*b b*)   The lagoon.

The subjoined cut (No. 7) exhibits a small part of the section of a coral island on a larger scale.

No. 7.

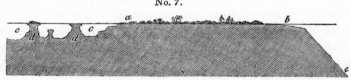

Section of part of a Coral Island.

(*a b*)   Habitable part of the island.
(*b e*)   Slope of the side of the island, plunging at an angle of forty-five to the depth of fifteen hundred feet.
(*c c*)   Part of the lagoon.
(*d d*)   Knolls of coral in the lagoon, with over-hanging masses of coral, resembling the capitals of columns.

The circular or oval forms of the numerous coral isles of the Pacific, with the lagoons in their centre, naturally suggest the idea that they are nothing more than the crests of submarine volcanos, having the rims and bottoms of their craters overgrown by corals.   This opinion is strengthened by the conical form of the submarine mountain, and the steep angle at which it plunges on all sides into the surrounding ocean. It is also well known that the Pacific is a great theatre of volcanic action, and every island yet examined in the wide region termed Eastern Oceanica, consists either of volcanic rocks or coral limestones.

It has also been observed that, although within the circular coral reefs, there is usually nothing discernible but a lagoon, the bottom of which is covered with coral, yet within some of these basins, as in Gambier's group, rocks composed of porous lava and other volcanic substances, rise up, resembling the two Kameni's, and other eminences of igneous origin, which have been thrown up within the times of history, in the midst of the Gulf of Santorin *.

We mentioned that in volcanic archipelagos there is generally one large habitual vent, and many smaller volcanos formed at different points and at irregular intervals, all of which have usually a linear arrangement. Now in several of the groups of Eastern Oceanica there appears to be a similar disposition, the great islands, such as Otaheite, Owhyhee, and Terra del Spirito Santo, being habitual vents, and the lines of small circular coral isles which are dependent on them being very probably trains of minor volcanos, which may have been in eruption singly and at irregular intervals.

The absence of circular groups in the West Indian seas, and the tropical parts of the Atlantic, where corals are numerous, has been adduced as an additional argument, inasmuch as volcanic vents, though existing in those regions, are very inferior in importance to those in the Pacific and Indian seas †. It may be objected that the circles formed by some coral reefs or groups of coral islets, varying as they do from ten to thirty miles and upwards in diameter, are so great as to preclude the idea of their being volcanic craters. In regard to this objection we may refer to what we have said in a former volume respecting the size of the so-called craters of elevation, many of which, we conceive, may be the ruins of truncated cones‡.

There is yet another phenomenon attending the circular reefs, to which we have not alluded, viz., the deep narrow passage which almost invariably leads from the sea into the lagoon, and is kept open by the efflux of the sea at low tides. It is sufficient that a reef should rise a few feet above low-water mark to cause the waters to collect in the lagoon at high tide,

* See vol. i. p. 386.
† De la Beche, Geol. Man. p. 141.   ‡ See vol. i. p. 388.

and, when the sea falls, to rush out violently at one or more
points where the reef happens to be lowest or weakest.   At
first there are probably many openings; but the growth of the
corals tends to obstruct all those which do not serve as the
principal channels of discharge, so that their number is gradually reduced to a few, and often finally to one.   This event
is strictly analogous to that witnessed in our estuaries, where
a body of salt-water accumulated during the flow, issues with
great velocity at the ebb of the tide, and scours out or keeps
open a deep passage through the bar, which is almost always
formed at the mouth of a river.

When we controverted in our first volume Von Buch's theory
of "elevation craters," we suggested that the single gorge leading from the central cavity to the sea, may have been produced
by a stream of water issuing from a lake filling the original
crater, and which had in process of time cut a deep channel*;
but we overlooked the more probable cause, the action of the
tides, which affords, we think, a most satisfactory explanation.
Suppose a volcanic cone, having a deep crater, to be at first
submarine, and to be then *gradually* elevated by earthquakes
in an ocean where tides prevail, a ravine cannot fail to be cut
like that which penetrates into the Caldera of the isle of Palma.
The opening would at first be made on that side where the rim
of the crater was originally lowest, and it would afterwards be
deepened as the island rose, so as always to descend somewhat
lower than the level of the sea.   Captain Beechey's observations,
therefore, of the effect of the tides on the coral islands, corro-
borate the opinion which we offered respecting the mode of formation of islands having a configuration like Palma; whereas
the theory of the *sudden* upheaving of horizontal strata into
a conical form, affords no explanation whatever of the *single*
ravine which intersects one side of these circular islands.

In the coral reefs surrounding those volcanic islands in the
Pacific which are large enough to feed small rivers, there is
generally an opening or channel opposite the point where the

* Vol. i., p. 395.

stream of fresh water enters the sea.   The depth of these channels rarely exceeds twenty-five feet, and they may be attributed, says Captain Beechey, to the aversion of the lithophytes to fresh water, and to the probable absence of the mineral matter of which they construct their habitations *.

But there is yet another peculiarity of the low coral islands, the explanation of which is by no means so obvious. They follow one general rule in having their windward side higher and more perfect than the other.   " At Gambier and Matilda islands this inequality is very conspicuous, the weather side of both being wooded, and of the former inhabited, while the other sides are from twenty to thirty feet under water, where, however, they might be perceived to be equally *narrow* and well defined.   It is on the leeward side also that the entrances into the lagoons occur ; and although they may sometimes be situated on a side that runs in the direction of the wind, as at Bow Island, yet there are none to windward." These observations of Captain Beechey accord perfectly with those which Captain Horsburgh and other hydrographers have made in regard to the coral islands of other seas.   Thus the Chagos Isles in the Indian Ocean are chiefly of a horse-shoe form, the openings being to the north-west ; whereas the prevailing wind blows regularly from the south-east.   From this fortunate circumstance ships can enter and sail out again with ease, whereas, if the narrow inlets were to windward, vessels which once entered might not succeed for months in making their way out again.   The well-known security of many of these harbours, depends entirely on this fortunate peculiarity in their structure.

In what manner is this singular conformation to be accounted for ?   The action of the waves is seen to be the cause of the superior elevation of some reefs on their windward sides, where sand and large masses of coral rock are thrown up by the breakers; but there are a variety of cases where this cause alone is inadequate to solve the problem ; for reefs

* Voyage to the Pacific, &c., p. 194.

submerged at considerable depths, where the movements of the sea cannot exert much power, have, nevertheless, the same conformation, the leeward being much lower than the windward side *.

I am informed by Captain King, that on examining the reefs called Rowley Shoals, which lie off the north-west coast of Australia, where the east and west monsoons prevail alternately, he found the open side of one crescent-shaped reef, the Impérieuse, turned to the east, and of another, the Mermaid, turned to the west; while a third oval reef, of the same group, was entirely submerged. This want of conformity is exactly what we should expect, where the winds vary periodically.

It seems impossible to refer the phenomenon now under consideration to any original uniformity in the configuration of submarine volcanos, on the summits of which we may suppose the coral reefs to grow ; for although it is very common for craters to be broken down on one side only, we cannot imagine any cause that should breach them all in the same direction. But, if we mistake not, the difficulty will be removed if we call in another part of the volcanic agency—subsidence by earthquakes. Suppose the windward barrier to have been raised by the mechanical action of the waves to the height of two or three yards above the wall on the leeward side, and then the whole island to sink down a few fathoms, the appearances described would then be presented by the submerged reef. A repetition of such operations by the alternate elevation and depression of the same mass (an hypothesis strictly conformable to analogy) might produce still greater inequality in the two sides, especially as the violent efflux of the tide has probably a strong tendency to check the accumulation of the more tender corals on the leeward reef, while the action of the breakers contributes to raise the windward barrier.

The calcareous formations of the Pacific are probably all stratified, although single beds may sometimes attain a great thickness. The occasional drifting of sand from the exposed

* Voyage to the Pacific, &c., p. 189.

parts of a reef into the lagoon or the surrounding sea, would suffice to form occasional lines of partition, especially during violent tempests which occur annually among the South-Sea islands. The decomposition of felspathic lavas may supply the current which washes and undermines the cliffs of some islands with fine clay, and this may be carried to great distances and deposited in distinct layers between calcareous masses, or may be mingled with them and form argillaceous limestones. Other divisions will arise from the arrangement of different species of testacea and zoophytes, which inhabit water of various depths, and which succeed each other as the sea deepens by the fall of the land during earthquakes, or grows shallower by elevation due to the same cause, or by the accumulation of organic substances raising the bottom.

To these causes of minor subdivision must be added another of great importance,—the ejection of volcanic ashes and sand, often carried by the wind over wide areas, and the flowing of horizontal sheets of lava, which may interrupt suddenly the growth of one coral reef, and afterwards serve as a foundation for another. An example of this kind is seen in the isle of France, where a bed of coral, ten feet thick, intervenes between two currents of lava *, and in the West Indies, in the island of Dominica, Maclure observes that " a bed of coral and madrepore limestone, with shells, lies horizontally on a bed of cinders, about two or three hundred feet above the level of the sea, at Rousseau, and is covered with cinders to a considerable height †."

The reefs in the Pacific are sometimes of great extent: thus the inhabitants of Disappointment Islands, and those of Duff's Group, pay visits to each other by passing over long lines of reefs from island to island, a distance of six hundred miles and upwards. When on their route they present the appearance of troops marching upon the surface of the ocean ‡.

* De la Beche, Geol. Man. p. 142.   Quoy and Gaimard, Ann. des Sci. Nat. tome vi.

† Observations on the Geology of the West Indian Islands, Journal of Science, &c., No. X., p. 318.   ‡ Malte-Brun's Geog. vol. iii. p. 401.

A reference to our first volume will show that a series of
ordinary earthquakes might, in the course of a few centuries,
convert such a tract of sea into dry land ; and it is, therefore,
a remarkable circumstance that there should be so immense an
area in eastern Oceanica, studded with minute islands, without
one single spot where there is a wider extent of land than belongs to such islands as Otaheite, Owhyhee, and a few others,
which either have been or are still the seats of active volcanos.
If an equilibrium only were maintained between the upheaving
and depressing force of earthquakes, large islands would very
soon be formed in the Pacific ; for, in that case, the growth of
limestone, the flowing of lava, and the ejection of volcanic
ashes, would combine with the upheaving force to form new
land.

Suppose the shoal which we have described as six hundred
miles in length, to sink fifteen feet, and then to remain un-
moved for a thousand years ; during that interval the growing coral may again approach the surface.   Then let the
mass be re-elevated fifteen feet, so that the original reef is
restored to its former position : in this case the new coral
formed since the first subsidence, will constitute an island
six hundred miles long.   An analogous result would have
occurred if a lava-current fifteen feet thick had overflowed the
submerged reef.   The absence, therefore, of more extensive
tracts of land in the Pacific seems to show that the amount of
subsidence by earthquakes exceeds in that quarter of the
globe at present the elevation due to the same cause.

We mentioned that one of the thirty-two islands examined
by our navigators in the late expedition, was raised about
eighty feet above the level of the sea*.   It is called Elizabeth
or Henderson's Island, and is five miles in length by one
in breadth.   It has a flat surface, and on all sides except the
north, is bounded by perpendicular cliffs about fifty feet high,
composed entirely of dead coral, more or less porous, honey-
combed at the surface, and hardening into a compact calcareous

* According to some accounts between sixty and seventy feet.

mass, which possesses the fracture of secondary limestone, and has a species of millepore interspersed through it. These cliffs are considerably undermined by the action of the waves,

No. 8.

Elizabeth or Henderson's Island.

No. 9.

Enlarged view of part of Elizabeth or Henderson's Island.

and some of them appear on the eve of precipitating their superincumbent weight into the sea. Those which are less injured in this way present no alternate ridges or indication of the different levels which the sea might have occupied at different periods, but a smooth surface, as if the island, which has probably been raised by volcanic agency, had been forced up by one great subterraneous convulsion *.

At the distance of a few hundred yards from this island, no bottom could be gained with two hundred fathoms of line. It will be seen from the annexed sketch, communicated to me by Lieutenant Smyth, of the Blossom, that the trees come down to the beach towards the centre of the isle, a break which at first sight resembles the openings which usually lead into lagoons: but the trees stand on a steep slope and no hollow of an ancient lagoon was perceived. The reader will remark that such a mass of limestone represents exactly those horizontal cappings of calcareous strata which we sometimes find on hills which have tabular summits.

As we have at present no proof that Henderson's Island has been upheaved within the historical period, we deviate somewhat from our plan when we describe it in the present chapter; but, as earthquakes are now felt from time to time in this part of the Pacific, and as indications of very recent changes of level

* Beechey, ib. p. 46.

are not wanting*, it is by no means improbable that the era of the elevation of this island may not be very remote.

The calcareous masses which we have now considered, constitute, together with the associated volcanic formations, the most extensive of the groups of rocks which can be demonstrated to be now in progress. The space in the sea which they occupy is so vast, that we may safely infer that they exceed in area any group of ancient rocks which can be proved to have been of contemporaneous origin. We grant that each of the great archipelagos of the Pacific are separated by unfathomable abysses, where no zoophytes may live and no lavas flow, where not even a particle of coral sand or volcanic scoriæ may be drifted : we confine our view to the extent of reef ascertained to exist, and assume that a certain space around each volcanic or coral isle has been covered with ejections or matter from the waste of cliffs, and it will then be seen that the space occupied by these formations may equal, and perhaps exceed in area that part of our continents which has been accurately explored by the geologist.

That the increase of these calcareous masses should be principally, if not entirely, confined to the shallower parts of the ocean, or, in other words, to the summits of submarine ranges of mountains and elevated platforms, is a circumstance of the highest interest to the geologist; for, if parts of the bed of such an ocean should be upraised, so as to form large continents, mountain-chains might appear, capped and flanked by calcareous strata of great thickness, and replete with organic remains, while in the intervening lower regions no rocks of contemporary origin would ever have existed.

A modern writer has attempted to revive the theory of some of the earlier geologists, that all limestones have originated in organized substances. If we examine, he says, the quantity of limestone in the primary strata, it will be found to bear a much smaller proportion to the siliceous and argillaceous rocks than in the secondary, and this may have some connexion with

* See Captain Beechey's Voyage to the Pacific, &c., pp. 159 and 191.

the rarity of testaceous animals in the ancient ocean. He farther infers that in consequence of the operations of animals, " the quantity of calcareous earth deposited in the form of mud or stone is always increasing; and that as the secondary series far exceeds the primary in this respect, so a third series may hereafter arise from the depths of the sea, which may exceed the last in the proportion of its calcareous strata *."

If these propositions went no farther than to suggest that every particle of lime that now enters into the crust of the globe, may possibly in its turn have been subservient to the purposes of life by entering into the composition of organized bodies, we should not deem the speculation improbable; but when it is hinted that lime may be an animal product combined by the powers of vitality from some simple elements, we can discover no sufficient grounds for such an hypothesis, and many facts which militate against it.

If a large pond be made, in almost any soil, and filled with rain water, it may usually become tenanted by testacea, for carbonate of lime is almost universally diffused in small quantities. But if no calcareous matter be supplied by waters flowing from the surrounding high grounds or by springs, no tufa or shell-marl are formed. The thin shells of one generation of molluscs decompose, so that their elements afford nutriment to the succeeding races; and it is only where a stream enters a lake, which may introduce a fresh supply of calcareous matter, or where the lake is fed by springs, that shells accumulate and form marl.

All the lakes in Forfarshire which have produced deposits of shell-marl, have been the sites of springs which still evolve much carbonic acid, and a small quantity of carbonate of lime. But there is no marl in Loch Fithie, near Forfar, where there are *no springs*, although that lake is surrounded by these calcareous deposits, and although, in every other respect, the site is favourable to the accumulation of aquatic testacea.

We find those charæ which secrete the largest quantity of

* Macculloch's Syst. of Geol., vol. i. p. 219.

calcareous matter in their stems, to abound near springs impregnated with carbonate of lime. We know that if the common hen be deprived altogether of calcareous nutriment, the shells of her eggs will become of too slight a consistency to protect the contents, and some birds eat chalk greedily during the breeding season.

If on the other hand we turn to the phenomena of inorganic nature, we observe that, in volcanic countries, there is an enormous evolution of carbonic acid, mixed with water or in a gaseous form, and that the springs of such districts are usually impregnated with carbonate of lime in great abundance. No one who has travelled in Tuscany, through the region of extinct volcanos and its confines, or who has seen the map recently constructed by Targioni to show the principal sites of mineral springs, can doubt for a moment, that, if this territory was submerged beneath the sea, it might supply materials for the most extensive coral reefs. The importance of these springs is not to be estimated by the magnitude of the rocks which they have thrown down on the slanting sides of hills, although of these alone large cities might be built, nor by a coating of travertin that covers the soil in some districts for miles in length. The greater part of the calcareous matter passes down in a state of solution to the sea; and a geologist might as well assume the mass of alluvium formed in a few years in the bed of the Po, or the Ganges, to be the measure of the quantity deposited in the course of centuries in the deltas of those rivers, as conceive that the influence of the carbonated springs in Italy can be estimated by the mass of tufa precipitated by them near their sources.

It is generally admitted that the abundance of carbonate of lime given out by springs, in regions where volcanic eruptions or earthquakes prevail, is referrible to the solvent power of carbonic acid. For, as the acidulous waters percolate calcareous strata, they take up a certain portion of lime and carry it up to the surface where, under diminished pressure in the atmosphere, it may be deposited, or, being absorbed by animals and vegetables, may be secreted by them. In Auvergne, springs charged with carbonate of lime rise through granite, in which case we must suppose the calcareous matter to be derived from some primary rock, unless we imagine it to rise up from the volcanic foci themselves.

We see no reason for supposing that the lime now on the surface, or in the crust of the earth, may not, as well as the silex, alumine, or any other mineral substance, have existed before the first organic beings were created, if it be assumed that the arrangement of the inorganic materials of our planet preceded in the order of time the introduction of the first organic inhabitants.

But if the carbonate of lime secreted by the testacea and corals of the Pacific, be chiefly derived *from below*, and if it be a very general effect of the action of subterranean heat to sub- tract calcareous matter from the *inferior* rocks, and to cause it to ascend to the surface, no argument can be derived in favour of the progressive increase of limestone from the magnitude of coral reefs, or the greater proportion of calcareous strata, in the more modern formations. A constant transfer of car- bonate of lime from the inferior parts of the earth's crust to its surface, would cause throughout all future time, and for an indefinite succession of geological epochs, a preponderance of calcareous matter in the newer, as contrasted with the older formations.

DESCRIPTION OF THE PLATES AND MAP.

View of part of the Valley del Bove, on the East side of the great Cone of Etna.

THIS valley is a cavity of immense depth, commencing at a short distance below the summit of Etna, and descending through that zone of the mountain where lateral eruptions are frequent. The general dip of the volcanic beds in the precipices surrounding this valley is towards the sea, but exceptions occur where lateral cones have been buried in the manner described in the first volume (p. 363). The stupendous precipices surrounding this great amphitheatre vary from 600 to nearly 3000 feet in height, and they are traversed on all sides by innumerable vertical walls or dikes of compact lava, which cut through the sloping beds of lava, sand, and scoriæ, of which the great cone is formed. These dikes, which will be described in the next volume, seem all to have been produced by ancient lateral eruptions on the flanks of Etna.

The causes which have produced this great depression in the otherwise symmetrical cone of the volcano will be discussed in the third volume, and we shall merely state here, that we consider the conformation of the rocky barrier encircling the cavity, as entirely at variance with an hypothesis recently proposed, that the hollow was a crater of eruption from whence the scoriæ of the surrounding heights have proceeded.

We have introduced two colours into the plate, the grey to express that part of the mountain which may have been formed before the origin of the " Val del Bove," the red to indicate the part which has resulted from eruptions subsequent to the formation of the valley. The great lava currents of 1819 and 1811, described in the first volume (p. 367), are seen pouring down from the higher parts of the valley, overrunning the forests of the great plain, and rising up in the foreground on the left with a rugged surface, on which small hillocks and depressions are seen, such as often characterize a lava-current immediately after its consolidation.

The small cone, No. 7, was formed in 1811, and was still smoking when I saw it in 1828. Immediately in front of it is seen another cone, formed during the same eruption. The other small volcano to the left, from which vapour is issuing, was formed, I believe, in 1819.

This sketch, which forms part of a panoramic drawing which I made in November 1828, is merely intended to assist the reader, in comprehending some geological details into which we shall hereafter enter, on the structure of the older portion of Etna, but it will give no idea of the extraordinary geological interest, still less of the picturesque grandeur of this magnificent scene of desolation. Nor is the view sufficiently extensive to exhibit the entire form of the vast amphitheatre, part only of the northern, and scarcely any of the southern boundary of which is included.

MAP

Shewing the extent of Surface in Europe which has been covered by Water since the Deposition of the older Tertiary Strata. (Strata of the Paris and London Basins, &c.)

[Constructed chiefly from M. Amie Boué's Geological Map of Europe.]

This map will enable the reader to perceive at a glance the great extent of change in the physical geography of Europe, which can be proved to have taken place since some of the older tertiary strata were deposited. The most ancient part of the period to which the map refers cannot be deemed

[In the original publication, the map on the following two pages was printed as a fold-out and hand painted in two colors. We have adapted the colors to shading in order to print the map in black and white.]

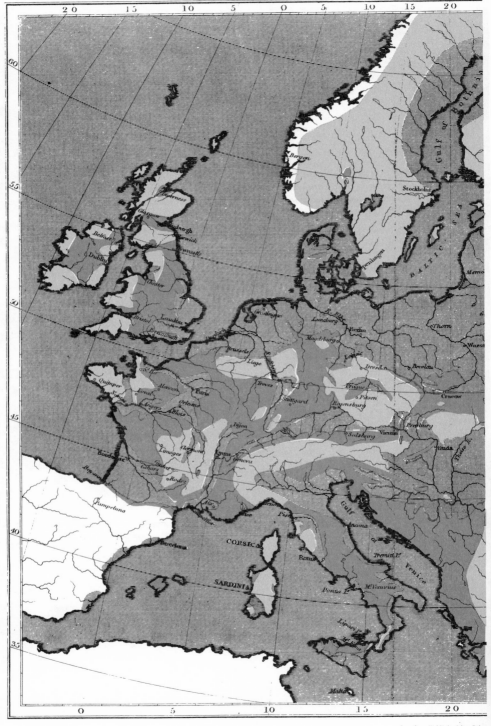

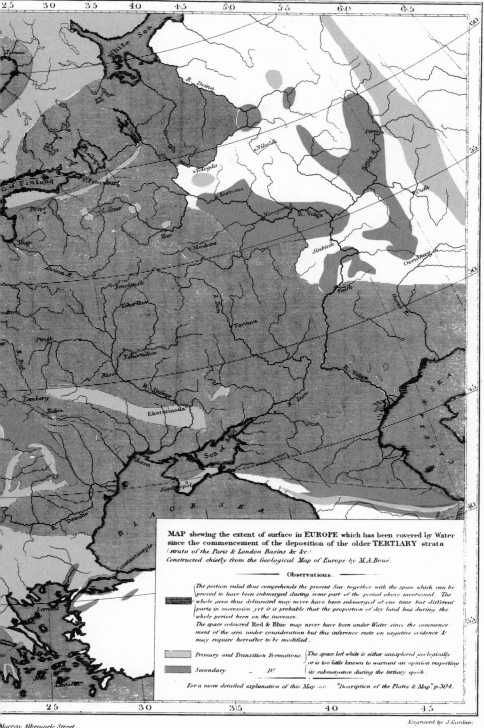

MAP shewing the extent of surface in EUROPE which has been covered by Water since the commencement of the deposition of the older TERTIARY strata (strata of the Paris & London Basins &c &c.) Constructed chiefly from the Geological Map of Europe by M.A.Boué.

Observations.

The portion ruled thus comprehends the present Sea, together with the space which can be proved to have been submerged during some part of the period above mentioned. The whole area thus delineated may never have been submerged at one time but different parts in succession yet it is probable that the proportion of dry land has during the whole period been on the increase.

The space coloured Red & Blue may never have been under Water since the commencement of the æra under consideration but this inference rests on negative evidence & may require hereafter to be modified.

Primary and Transition Formations.

Secondary   Dᵒ.

The space left white is either unexplored geologically or is too little known to warrant an opinion respecting its submergence during the tertiary epoch.

For a more detailed explanation of this Map see "Description of the Plates & Map" p.304.

Murray, Albemarle Street.

Engraved by J.Gardner.

very remote, considered geologically, because the deposits of the Paris and London basins, of Auvergne, and many other districts belonging to the older tertiary epoch, are newer than the greater part of the sedimentary rocks of which the crust of the globe is composed. The species, moreover, of marine and fresh-water testacea, of which the remains are found in these formations, are not entirely distinct from such as now live; a proportion of about three in a hundred of the fossil species having been identified with species now living. Yet, notwithstanding the comparatively recent epoch to which the retrospect is carried, the variations in the distribution of land and sea depicted on the map, form only a part of those which must have taken place during the period under consideration. Some approximation has merely been made to a correct estimate of the amount of *sea converted into land* in that part of Europe best known to geologists, but we cannot determine how much land has become sea during the same period; and there may have been repeated interchanges of land and water in the same places, mutations of which no account is taken in the map, and respecting the amount of which little accurate information can ever be obtained by geologists.

The proofs of submergence, during some part of the tertiary period, throughout the districts distinguished by ruled lines, are of a most unequivocal character; for the area thus described is now covered by deposits, containing the remains of aquatic animals belonging to tertiary species. We have, indeed, extended the sea in two or three instances beyond these limits, because other geological data have been obtained for inferring the submergence of these tracts subsequently to the commencement of the deposition of the tertiary strata. Thus we shall explain, in the next volume, our reasons for concluding that part of the chalk of England, (the north and south downs, for example, together with some other adjoining secondary tracts,) continued beneath the sea until the older tertiary beds had begun to accumulate.

It is possible also that a considerable part of Caernarvonshire might with propriety have been represented as sea, if our information respecting the geology of that country had been more full and accurate; for marine shells have been found in sand and gravel at the height of one thousand feet above the level of the sea, on the summit of Moel Tryfane, between Snowdon and the Menai Straits. The species are apparently recent, but certainly are newer than the older tertiary epoch *.

The introduction of a small bay where the river Ribble enters into the sea in Lancashire, is warranted by the newly discovered deposit of tertiary shells covering an area of about thirty miles square in that region †.

A portion also of the primary district in Brittany is divided into islands, because it has been long known to be covered with patches of marine tertiary strata; and when I examined the disposition of these, in company with my friend Captain S. E. Cook, R.N., in 1830, I was convinced that the sea must have covered much larger areas than are now occupied by these small and detached deposits.

The former connexion of the White Sea and the Gulf of Finland is proved by the fact that a broad band of tertiary strata extends throughout part of the intervening space. We have represented the channel as somewhat broader than the tract now occupied by the tertiary formation, because the latter is bordered on the north-west by a part of Finland, which is extremely low, and so thickly interspersed with lakes as to be nearly half covered with fresh-water.

Certain portions of the north-western shores of Norway have been left blank, because the discovery by Von Buch, Brongniart, and others, of deposits of recent shells along the coast of Norway and Sweden, at several places and at various heights above the level of the sea, attest the comparatively

* Joshua Trimmer, Esq., Proceedings of the Geological Society of London, No. 22, 1831. The shells were exhibited at the Geological Society when the memoir was read.

† See an abstract of a memoir read by Mr. Murchison, Pres. Geol. Soc., Proceedings of York meeting, 1831.

recent date of the elevation of part of the gneiss and other primary rocks in that country, although we are unable as yet to determine how far the sea may have extended.

On the other hand, a considerable space of low land along the shores of the Gulf of Bothnia, in the Baltic, is represented as sea, because the growth of deltas on that coast, and the shallowing of the water by sedimentary deposits during the historical era, leave no room for doubt that the extent of the gulf must have been very much greater at some periods since the older tertiary epoch.

The low granitic steppe coloured red, to the north of the Black Sea has *not* been represented as having been under water during the tertiary period, although, from the quantity of marine tertiary strata in the surrounding districts, it is far from improbable that it has recently emerged.

We were anxious, in the observations annexed to the title of this map, to guard the reader against the supposition that it was intended to represent the state of the physical geography of part of Europe at any one period. It is not a restoration of a former condition of things, but a view of the change which a certain amount of surface has undergone within a given period, an alteration so complete, that not one of the species of organic beings which now inhabit the large space designated by ruled lines, beyond the borders of the existing seas, can have lived there during some other period subsequent to the commencement of the tertiary era.

We have stated, in the first volume *, that the movements of earthquakes occasion the subsidence as well as the upraising of the surface; and that, by the alternate rising and sinking of particular spaces, at successive periods, a great area may have been entirely covered with marine deposits, although the whole may never have been beneath the waters at one time; nay, even though the relative proportion of land and sea may have continued unaltered throughout the whole period. We believe, however, that since the commencement of the tertiary

* Page 126.

X 2

period, the dry land in the northern hemisphere has been continually on the increase, not only because it is now greatly in excess beyond the average proportion which land generally bears to water on the globe, but because the comparison of the secondary and tertiary strata implies a passage throughout the space now occupied by Europe, from the condition of an ocean interspersed with islands to that of a large continent.

But if it were possible to represent all the vicissitudes in the distribution of land and sea that have occurred during the tertiary period, and to exhibit not only the actual existence of land where there was once sea, but also the extent of surface now submerged, which may once have been land, the map would still fail to express all the important revolutions in physical geography, which have taken place within the epoch under consideration. The oscillations of level have not merely been such as to lift up the land from below the waters to a small height above them, but in some cases a rise of several thousand feet has been effected. Thus the Alps have acquired an additional altitude of from 2000 to 4000 feet, and even in some places still more; and the Apennines owe a great part of their height (from 1000 to 2000 feet and upwards) to subterranean convulsions which have happened within the tertiary epoch.

On the other hand, some mountain chains may have been lowered, during the same series of ages, in an equal degree, and shoals may have been converted into deep abysses.

It would be superfluous to point out in detail the bearing of the facts exhibited in this map, on the theories proposed in a former part of this volume, respecting the migrations of animals and plants, and the extinction of species; and it would be equally unnecessary to enlarge on the variations in *local* climate, which must have accompanied such vicissitudes in physical geography.

But the general temperature, also, of the habitable surface of the globe, as well as the local climates, may have been considerably modified by such extraordinary revolutions. The alteration in climate, implied by a comparison of the organic remains of the older tertiary strata, and the species of living animals and plants, does not appear to be so great as would be produced if the temperature of our tropics were now transferred to the temperate zone, and the temperature of the latter to the arctic. We do not, therefore, anticipate that the reader, who has duly studied the arguments explained by us in the 6th, 7th, and 8th chapters of the first volume, will object to the *adequacy* of the cause proposed, on the score of the small quantity of geographical change during the time in question.

But if there be good reason to conclude that the change would be fully adequate, in point of the magnitude of its effects, this cause, we conceive, ought to supersede every other of a purely speculative nature, until some argument can be adduced to prove that the change has not acted in the right direction *.

Some persons, but slightly acquainted with the present state of geology, have objected, that the lands in high northern latitudes have *not* been recently elevated. If they had reflected that every year we are making some new discoveries respecting the periods when tracts in the immediate neighbourhood of the great European capitals emerged from the deep, and had they sufficiently considered that the antiquity of a group of rocks has no necessary connexion with the date of its elevation, they would probably have seen the futility of such arguments. As far as we can conjecture, from the very scanty information which we possess of the geology of the arctic region, there is no want of proofs of comparatively recent alterations of level.

In conclusion, we may remark that the portion of Europe distinguished in this map by colours and ruled lines, comprises the greater part of the globe now known to geologists—almost all at least that is known in such a manner as to entitle any one to speculate on the mutations in physical geography which have taken place during the tertiary period.

* See Mr. Herschell's remarks on a change of climate.—Disc. on the Study of Nat. Phil., pp. 146 and 148.

In regard to other parts of the world, we have no reason for inferring, from any data hitherto obtained, that during an equal lapse of the ages which immediately preceded our times, an equal amount of alteration of surface may not have taken place.

LIST OF WOOD-CUTS.

INDEX.

VOL. II.

END OF THE SECOND VOLUME.

LONDON:
PRINTED BY WILLIAM CLOWES,
Stamford Street.